High Strain, High Stress: The Challenge of Modeling Cavitation in Biomechanics

Naviya

CONTENTS

Chapter 1

INTRODUCTION

1.1 Motivation

In biomedical sciences, numerical and modeling approaches to study biological systems often present significant challenges due to their multiscale nature and the interaction of various materials ranging from liquids and soft tissue to elastic solids. Many applications in the medical field involve the—often nonlinear—dynamical behavior of these materials in high strain-rate and/or high stress regimes. One example where such strain-rates and stresses are generated is during cavitation of single or clouds of bubbles. Treatments making use of focused ultrasound, shock waves, or laser surgery can cause nucleation of bubble clouds which collapse rapidly, deforming the surrounding material at high rates and inducing very high pressures at the bubble nucleation site. Understanding the interaction of such cavitating gas bubbles with viscoelastic materials and stiff elastic solids is thus valuable in various applications in the biomedical field.

In this book, we present numerical and data-driven methods for the modeling of multiphase physics involving cavitating bubble dynamics in and near elastic and viscoelastic materials. In the first part, we make use of data assimilation with bubbles cavitating in a viscoelastic hydrogel to estimate its mechanical properties. This method to characterize soft materials can aid in the modeling of biomaterials in settings where high strain rates occur. In the second part, we implement a numerical framework for the direct numerical simulation of acoustic wave–bubble–stone interactions. This is used to simulate burst-wave lithotripsy, a treatment to break kidney stones through exposure to repeated bursts of focused ultrasound. In particular, this framework enables a better understanding of the impact and dynamics of bubbles that can form near the stone during treatment.

1.2 Inertial Microcavitation Rheometry

Across various engineering disciplines, particularly in medicine and the biomedical field, knowing the mechanical properties of soft materials such as hydrogels, polymers, and biomaterials is of great importance (Chaudhuri et al., 2016; Lee and Mooney, 2012; Storrie and Mooney, 2006; Solomon and Jindal, 2007). This includes applications where high strain rates occur in biological tissue, for example during impact and blast exposure (Bar-Kochba et al., 2016; Sarntinoranont et al., 2012; Meaney and Smith, 2011; Nyein et al., 2010; Ramasamy et al., 2011), therapeutic ultrasound (Maxwell, Cain, Duryea, et al., 2009; Xu et al., 2007; Mancia, Vlaisavljevich, Xu, et al., 2017; Mancia, Vlaisavljevich, Yousefi, et al., 2019; Vlaisavljevich, Lin, Warnez, et al., 2015; Bailey, Khokhlova, et al., 2003) or laser surgery (Brujan and Vogel, 2006; Vogel et al., 2008). However, measuring the mechanical properties of soft, viscoelastic materials at these high strain rates (exceeding $10^3 \, \mathrm{s}^{-1}$)) is challenging, in particular due to the high compliance of these materials (Arora, Narani, and McCulloch, 1999) and the strain rate dependence of their properties (Brujan and Vogel, 2006). At high strain rates, impact (Taylor, 1948; Allen, Rule, and Jones, 1997) or Kolsky bar tests (Chen and Song, 2010) can be used, but neither is adequate for soft materials, as they rely on high loading rates and applied stresses, incompatible with the high compliance of viscoelastic materials of interest (Hu, Zhao, et al., 2010; Hu, You, et al., 2012). Thus, measuring the mechanical properties of soft, viscoelastic materials at high strain rates remains a challenging goal of rheometry.

To this end, methods relying on cavitation in these soft materials have been developed. The bubble dynamics that occur when a bubble collapses in a soft material are sensitive to the material's properties, and thus observation of these dynamics can be exploited to characterize these properties. The first such method, called the Cavitation Rheology Technique (CRT), was introduced by Zimberlin, Sanabria-DeLong, et al. (2007). They make use of needle-induced cavitation, where a cavity is created in a gel at the tip of a needle by introducing pressurized air. The elastic modulus of the gel is deduced by determining the critical pressure where rapid deformation of the gel occurs. This technique has been studied extensively, and used, for example, to determine the mechanical properties of polymers and biomaterials in the eye and skin (Barney et al., 2020; Zimberlin, Sanabria-DeLong, et al., 2007; Zimberlin, McManus, and Crosby, 2010; Cui et al., 2011; Chin et al., 2013; Bentz, Walley, and Savin, 2016). An extension of this method, which does not rely on determining the critical pressure, but instead tracking the volume expansion of the cavity, was introduced by Raayai-Ardakani, Chen, et al. (2019) and Raayai-Ardakani and Cohen (2019) and used to characterize biological tissue (Mijailovic et al., 2021; Van

Sligtenhorst, Cronin, and Brodland, 2006). While effective for smaller strain rates, these methods cannot reach the high strain rate regime of interest to the biomedical applications described above (Chockalingam et al., 2021).

To this end, Estrada et al. (2018) proposed a high-strain rate rheometer to estimate the viscoelastic properties of polyacrylamide gels through observation of the bubble radius time history during a laser-generated cavitation event in a sample of the material. Cavitation occurring in liquids and soft materials on exposure to tensile waves of sufficient amplitude will induce very high strain rates in the material, within the regime of interest. The technique, called Inertial Microcavitation Rheometry (IMR), compares bubble radius time-histories obtained through high-speed imaging of laser-induced cavitation (LIC) in the material of interest, to simulated radius vs. time curves. Using an adequate theoretical cavitation model derived from literature on cavitation in a fluid (Akhatov et al., 2001; Epstein and Keller, 1972; Flynn, 1975; Fujikawa and Akamatsu, 1980; Keller and Kolodner, 1956; Keller and Miksis, 1980; Nigmatulin, Khabeev, and Nagiev, 1981; Prosperetti, 1991; Prosperetti, Crum, and Commander, 1988; Prosperetti and Lezzi, 1986), an array of simulations are run with varying material properties and the difference with experimental data is minimized through least-squares fitting to determine parameters which best match the experiment. In this way, properties of the material can be inferred through observation of these cavitating bubbles. Estrada et al. (2018) demonstrate that they can adequately infer the shear modulus and viscosity of polyacrylamide gels with varying stiffness using this method. This technique can achieve a broad range of strain rates, from $O(10^3)s^{-1}$ to $O(10^8)s^{-1}$. It is the first reliable technique for rheometry of soft materials in this regime. The method is minimally invasive and the setup relatively straightforward, making it broadly applicable to an array of gels or other soft materials with high compliance. The experimental setup, as well as representative bubble images captured by the high-speed camera which are used to determine bubble radius time-histories are shown in figure 1.1.

Results obtained in well-characterized polyacrylamide hydrogels by Estrada et al. (2018) are promising. However, there are a few areas where their method could be improved. First, the computational cost of the least-squares fitting approach used is not easily scalable, particularly to the estimation of additional parameters, as may be necessary given other material models. Only two parameters (shear modulus and viscosity) are estimated by Estrada et al. (2018), which already required N^2 simulations, with N the number of guesses per parameter. Unless a good estimate

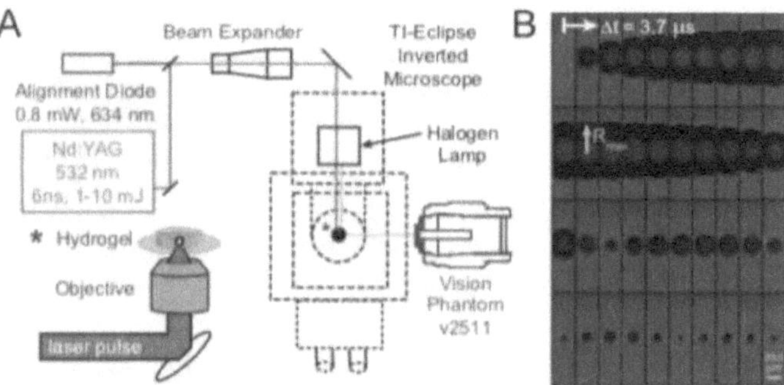

Figure 1.1: Sketch of the IMR experimental setup (A) and time-history of representative captured bubble images (B). Reprinted from Estrada et al. (2018) with permission from Elsevier, © 2018.

for each parameter is known apriori, N must be relatively large to obtain a precise estimate, with its size growing by an order of magnitude with each added significant figure. If material models require a third parameter to estimate, the number of simulations rises to N^3, and so on for each additional parameter. In the case tested where only two parameters are estimated, in a material that has been extensively studied, this is tractable. However, scaling to other methods may prove difficult.

Second, the reliance of IMR on laser-induced cavitation introduces some uncertainty in the parameter estimation. The physics of laser nucleation are not captured in the model used, particularly in the initial growth phase. Thus, Estrada et al. (2018) fit radius curves starting at the maximum bubble radius. This is a reasonable approach to avoid the complicated growth phase, and some uncertainty cannot be avoided unless a new bubble-dynamics model is used that captures plasma generation by the laser. However, IMR would benefit from a framework that better quantifies this uncertainty during parameter estimation and provides additional information as to the expected error in obtained estimates.

1.3   Data Assimilation

The fundamental idea behind data assimilation (DA) is to combine a mathematical model for a phenomenon with data, or observations, in order to best determine its state and predict its behavior. In many cases, such as those explored in this book, we begin with a forward model for the system. That is, we derive a mathematical model which, given a system at time t, will predict its state at time $t + 1$. This model can be analytical, or a simulation tool providing an approximate solution. At each time when data (experimental or otherwise) is available, it is used to correct and improve the state prediction. This is often done sequentially, where the model forecast and data-informed correction are applied in turn at each time, but can also be used to determine the initial state of a system and estimate particular parameters. Data assimilation can thus be considered a framework to solve inverse problems, where data is combined with a mathematical model to estimate unknown parameters in the initial condition of a dynamical system (Sanz-Alonso, Stuart, and Taeb, 2023; Law, Stuart, and Zygalakis, 2015; Schillings and Stuart, 2017). Most DA methods are designed to account for uncertainty in both the model and data, which makes them versatile and adaptable to contexts with noisy data or using reduced-order models.

Applications of data assimilation are plentiful and span many disciplines. One example is the state estimation for guidance, navigation and control of aircraft, spacecraft, and other vehicles. Here, observation data from various onboard instruments can be combined online with modeled attitude and position for precise determination of the state of the craft or vehicle (Lefferts, Markley, and Shuster, 1982; Kim et al., 2007; Madyastha et al., 2011; Venhovens and Naab, 1999; Wenzel et al., 2006; Antonov, Fehn, and Kugi, 2011). Another is atmospheric and oceanographic modeling, for example for weather forecasting where sparse data points (e.g., pressure, temperature, and wind measurements) are combined with fluid dynamics models to best predict weather and atmospheric conditions. This problem is a great example of how useful DA can be. Indeed, precise direct modeling of weather conditions on an atmospheric scale is not feasible given current (or near-future) computing power. On the other hand, only limited data points can be obtained at weather tracking stations or using weather balloons. Neither modeling tools nor data alone are sufficient to accurately predict future weather conditions. However, combining reduced-order fluid dynamical models with sparse atmospheric measurements in a data assimilation framework is a powerful tool for atmospheric modeling and weather prediction. In fact, many modern DA methods have emerged from atmospheric sciences and weather forecasting, where it is widely used (Houtekamer and Zhang, 2016; Leeuwen, 2009; Lorenc, 1986; Bengtsson, Snyder, and Nychka, 2003; Evensen and Van Leeuwen, 1996). The DA methods used in this book are all extensions or variations on perhaps the most well-known and widely used DA method, which was foundational to the field of data assimilation: the Kalman filter (Kalman, 1960; Kalman and Bucy, 1961).

The Kalman Filter and Kalman Smoother To describe the Kalman filter, we begin by defining the linear dynamics model

$$x_{k+1} = Fx_k + \eta_k \tag{1.1}$$

where $F \in \mathbb{R}^{d \times d}$ is a linear operator for the state transition model, which maps a state $x \in \mathbb{R}^d$ from time k to $k + 1$, and $\eta_k \sim \mathcal{N}(0, \Sigma)$ is the model error, assumed Gaussian with mean 0 and variance Σ. The state is viewed as a random variable $x_k \sim \mathcal{N}(m_k, C_k)$ with mean m_k and covariance C_k. The estimate of the state at a time k is thus the mean of this random variable x_k. The state transition model F is generally based on a mathematical model for the evolution of the dynamical system, and must, in this case, be linear to be represented in matrix form.

We also define a linear observation model

$$y_{k+1} = Hx_{k+1} + v_{k+1}, \tag{1.2}$$

where $H \in \mathbb{R}^{n \times d}$ is the linear observation operator, which maps the state x_{k+1} to measurement space, where $y_{k+1} \in \mathbb{R}^n$ is the observation (data) at time $k + 1$. $v_k \sim \mathcal{N}(0, \Gamma)$ is the measurement noise, again assumed Gaussian with mean 0 and variance Γ. The Kalman filtering process, which estimates the optimal (estimation variance minimizing) state at step $k + 1$ given the dynamical system described by equations (1.1) and (1.2) can be broken down into a forecast step and an analysis step. We track the random Gaussian state variable x_k via its mean m_k and covariance C_k at any given time k. The mean and covariance are forecast and updated using the following. We note that only the final formulas used in the Kalman filter are reported here. A clear and thorough example of their full derivation can be found in Sanz-Alonso, Stuart, and Taeb (2023).

Forecast Step Since the state is represented by a Gaussian random variable (which is independent of the model error noise η, and the forecast operator is linear, we can write

$$\widehat{m}_{k+1} = Fm_k \tag{1.3}$$

$$\widehat{C}_{k+1} = FC_kF^T + \Sigma, \tag{1.4}$$

$$\tag{1.5}$$

where $\widehat{\cdot}$ represents forecast variables. Here, we simply propagate the mean and covariance of our state through the forecast operator.

Analysis Step   Once the forecast step is done, we introduce our measurement y_{k+1} to correct the forecast state statistics. This yields

$$m_{k+1} = \widehat{m}_{k+1} + K_{k+1}(y_{k+1} - H\widehat{m}_{k+1}) \tag{1.6}$$

$$C_{k+1} = (I - K_{k+1}H)\widehat{C}_{k+1}, \tag{1.7}$$

where

$$K_{k+1} = \widehat{C}_{k+1}H^T(H\widehat{C}_{k+1}H^T + \Gamma)^{-1} \tag{1.8}$$

is called the *Kalman gain*, and the vector $y_{k+1} - H\widehat{m}_{k+1}$ in equation (1.6) is called the *innovation*. The Analysis step can also be re-framed from an optimization perspective, which will be particularly useful when comparing it to other DA methods. We can re-write this analysis step as the minimization of a cost function J_{KF}:

$$m_{k+1} = \text{argmin}_x J_{KF}(x), \tag{1.9}$$

where

$$J_{KF}(x) = \frac{1}{2}\|y_{k+1} - Hx\|^2_\Gamma + \frac{1}{2}\|x - \widehat{m}_{k+1}\|^2_{\widehat{C}_{k+1}}. \tag{1.10}$$

This cost function helps understand the Kalman filter analysis intuitively. The first term minimizes the difference between the state vector mapped to measurement space and the observation, weighed by the measurement noise covariance. The second term minimizes the difference between the state vector with the projected state vector given the physical model, weighed by the model error covariance.

Another class of DA methods are smoothers. The difference between *filtering* and *smoothing* is that in filtering, data is incorporated into the state estimate sequentially at each time step. In this sense, it is often referred to as an 'online' method, as the estimate is adjusted at each time. By contrast, smoothing is done 'offline,' incorporating data from multiple time steps to adjust the state estimate. In filtering, only data from time k is used to estimate x_k. In smoothing, data from multiple times (e.g., k to $k+n$) is used to estimate this same state. The Kalman smoother is a simple extension of the Kalman filter, best understood in the optimization framework, where the cost function is modified to account for data at multiple times, yielding

$$J_{KS}(X) = \frac{1}{2}\|x_0 - m_0\|^2_{C_0} + \frac{1}{2}\sum_{k=0}^{N-1}\|y_{k+1} - Hx_{k+1}\|^2_{\Gamma} + \frac{1}{2}\sum_{k=0}^{N-1}\|x_{k+1} - Fx_k\|^2_{\widehat{C}_{k+1}}, \tag{1.11}$$

where $X = \{x_0, ..., x_N\}$ represents the time-series of the state from time 0 to N. This cost function minimizes once again the difference with both data and forecast state, but this time over a series of times 0 to N, and the state at all times is estimated.

1.4 The DA-IMR Approach

We propose a data assimilation approach to inertial microcavitation rheometry. The goal of IMR, namely estimation of the mechanical properties of a viscoelastic material by observation of bubble dynamics, is posed as an inverse problem that is solved using data assimilation. In this framework, we require a bubble dynamics solver for our operator F in the dynamical model given in equation (1.1), and the state x will be comprised of all dependent variables of this solver. The data y is simply the bubble-radius observations, and the observation operator H maps this observed radius to the state space. The key to using data assimilation to estimate material properties is then simply to append these unknown properties to the state vector x, with an initial estimate at initial time for x_0. As long as the dynamics of the bubble radius evolution are sensitive to changes in these properties, they will be estimated at each step in the filtering. This represents the basis of DA-IMR, our data assimilation approach to inertial microcavitation rheometry.

The goal of DA-IMR is twofold. First, this method can vastly improve the efficiency of IMR. IMR requires least-squares fitting of a large number of curves to experimental data and thus requires N^i bubble collapse simulations, where i is the number of parameters to estimate and N the number of values to test per parameter. This quickly becomes intractable in order to achieve large precision or to estimate many parameters. Second, DA can provide interesting insight into both modeling and experimental error in the estimation process. Tracking covariances associated with the forecast and analysis steps can provide insight into both model uncertainty and experimental error which could be present. Furthermore, the ability of filters to estimate parameters online means that we can obtain time-dependent information about this uncertainty.

However, an issue appears in order to use Kalman filtering (KF) or smoothing (KS) for this application. As described in section 1.3, the operators F and H used in the forecast and analysis steps of the KF and KS must be *linear*. This poses no issue for the observation operator H here, as the radius is a dependent variable of our model, and thus H will simply be a one-to-one map of the radius to itself. To capture bubble dynamics in a viscoelastic material, however, our dynamical model will need to be nonlinear. Indeed, capturing these dynamics with a sufficient level of accuracy usually requires solving a system of at least 2 nonlinear ordinary differential equations (for bubble radius and pressure), and 2 partial differential equations (for temperature and vapor concentration in the bubble). These models will be described in Chapter 2, but the issue in the context of Kalman filtering is their nonlinearity, which means they cannot be represented by an operator F in matrix form. This means a different approach is required to update the Covariance matrix at each step. A simple approach to deal with weakly nonlinear systems is the so-called extended Kalman filter (Jazwinski, 2007; Gelb, 1974; Ghil et al., 1981). This method simply linearizes the dynamics about the mean, using the Jacobian matrix to represent F and calculate the Kalman gain. However, this method is inadequate for highly nonlinear dynamics, as is the case of our bubble dynamics system (Julier and Uhlmann, 2004).

Instead, we turn to a class of methods called ensemble methods, which are widely used in problems with high-dimensional systems and nonlinear dynamics. The driving idea behind ensemble methods is to represent the statistics of the state empirically by randomly sampling q realizations of the state from a given distribution. We thus create an ensemble, the statistics of which can be empirically determined. The mean and covariance of this ensemble are used as a surrogate measure of the state estimate and error covariance. This removes the need to update the covariance with a linear forecast operator, as the covariance is simply calculated empirically from the ensemble, each element of which can be updated with any dynamical model. The particular methods used in DA-IMR, which will be detailed in Chapter 2, are Ensemble Kalman methods, which are ensemble-based extensions of the Kalman filter and smoother described in section 1.3, making use of this empirical sampling (Evensen, 1994; Evensen, 2003; Evensen, 2004; Houtekamer and Derome, 1995; Houtekamer and Mitchell, 1998; Anderson and Anderson, 1999).

1.5 Burst-Wave Lithotripsy

The second part of this book, in Chapters 4 and 5, focuses on a different biomedical application: the treatment of kidney stone disease with lithotripsy. Renal calculi occur in millions of patients around the world every year, with the number of individuals affected in the United States rising above 10% of the population in the past decade (Scales et al., 2012; Hill et al., 2022). Various treatments exist to break and remove kidney and urinary stones. While surgical removal is effective, it presents complication risks, is expensive, and can be difficult depending on the stone location. Instead, an effective and safe non-invasive treatment can create better outcomes for patients. Extracorporeal Shock-wave lithotripsy (SWL) is a common, non-invasive treatment for kidney stone disease. In SWL stones are subject to multiple rounds of focused shock waves to break them into smaller pieces that can then be passed naturally (Chaussy, Brendel, and Schmiedt, 1980; Chaussy, Schmiedt, et al., 1982; Chaussy and Fuchs, 1989; Sackmann et al., 1988). SWL is applied non-invasively under local or general anesthesia to avoid pain for the patient, and $O(1000)$ shocks are delivered at a rate of approximately 1 Hz. Various studies have examined treatment parameters such as shock wave rate, lithotripter technology, and shock wave amplitude variation (Lingeman et al., 2009; Jendeberg et al., 2017).

While SWL avoids the need for surgery and is generally effective, the shock waves generated have very high pressure magnitudes, which can cause damage to surrounding tissue, internal bleeding, and other adverse effects in patients (Evan et al., 1998; McAteer and Evan, 2008; Bailey, Pishchalnikov, et al., 2005; Janetschek et al., 1997). Burst-wave lithotripsy (BWL) was developed as an alternative, to improve comminution and mitigate risks associated with SWL. Instead of shock waves, BWL uses repeated short bursts of ultrasound (Maxwell, Cunitz, et al., 2015). The frequency is $O(100)$ kHz, with pulses of 10 to 20 cycles of ultrasound occurring at a pulse repetition frequency (PRF) of 1 to 10 Hz. In BWL, peak positive pressures are greatly reduced compared to SWL, where pressure magnitudes can reach up to 100 MPa. During treatment, this peak pressure is generally limited so that negative pressure magnitudes need not exceed 7 MPa to minimize the risk of cavitation-induced injury. An added advantage is that BWL does not require anesthesia, as patients tolerate the procedure well (Harper, Metzler, et al., 2021). Figure 1.2 compares the SWL and BWL waveforms, highlighting the difference in pressure amplitude between the treatments.

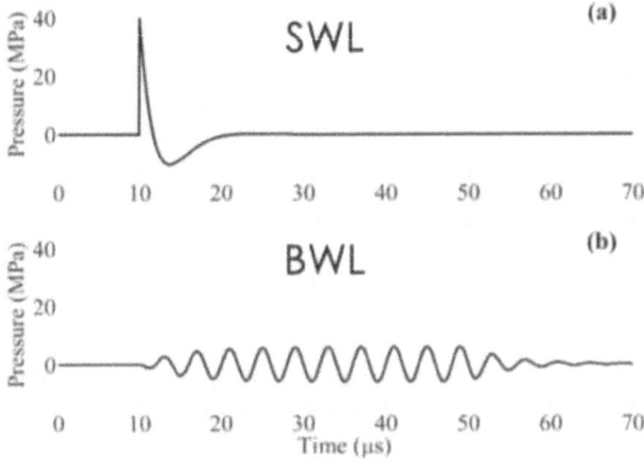

Figure 1.2: Comparison of typical SWL (top) and BWL (bottom) focal pressure waveforms. Reprinted from Maxwell, Cunitz, et al. (2015) with permission from Wolters Kluwer Health, © 2015.

Experimental studies of BWL, both in-vitro (Maxwell, Cunitz, et al., 2015) and in-vivo in a pig model (Maxwell, Wang, et al., 2019), have shown that it can break stones effectively. Maxwell, Cunitz, et al. (2015) fractured stones of various compositions, including stones known to be difficult to break with SWL, and compared treatment efficacy at various frequencies. Notably, they observed the comminution of stones into smaller fragments at higher frequencies. Compared to SWL, BWL thus leads to a more uniform fragment size that can be controlled by ultrasound frequency. A schematic of the experimental setup used by Maxwell, Cunitz, et al. (2015), taken from their paper, is shown in figure 1.3. Sapozhnikov, Maxwell, and Bailey (2021) and Bailey, Maxwell, et al. (2022) have shown that higher frequencies of ultrasound can enhance stone comminution for smaller calculi, further pointing to the fact that this frequency-dependence may be exploited to adapt BWL to stone size. More recently, clinical trials have shown promise in treating human patients (Harper, Metzler, et al., 2021). In a trial on 19 patients, Harper, Lingeman, et al. (2022) successfully fragmented 91 % of treated stones, achieving a median stone comminution of 90 % of stone volume.

These results show that BWL can be very effective. Studies have looked at the structure of elastic waves in stones during treatment experimentally (Maxwell, MacConaghy, et al., 2020), and developed models to predict maximum stress in

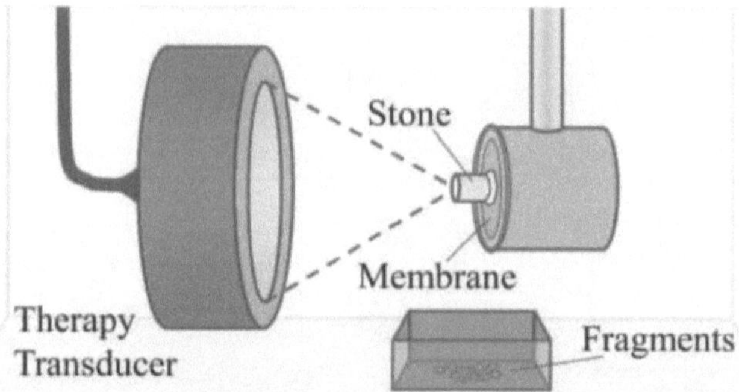

Figure 1.3: Schematic of the experimental setup for in-vitro BWL used by Maxwell, Cunitz, et al. (2015) and representative of the setup used for most experimental studies of BWL. Reprinted from Maxwell, Cunitz, et al. (2015) with permission from Wolters Kluwer Health, © 2015.

the stone in idealized settings (Sapozhnikov, Maxwell, and Bailey, 2021). However, as a novel treatment, many questions remain as to the mechanisms of stone fragmentation during BWL. A complicating factor is that high pressure magnitudes near the proximal surface of the stone can induce cavitation, resulting in bubble clouds forming between the transducer and the stone.

1.6 Bubble Formation during BWL

As mentioned, a remaining challenge of BWL is the propensity for cavitation to occur, particularly near the stone's proximal surface. In SWL it is generally accepted that cavitation, while undesirable from the point of view of injury, is important for complete stone comminution. While stress waves generated by the focused shock alone tend to result in large fragments, bubble-collapse-induced erosion produces much smaller ones. Inter-pulse times of 0.5 to 1s in SWL ensure that cavitation bubbles induced in each pulse are largely dissolved by the following one

Some approaches have tried to exploit this stone-erosion potential of bubble cloud collapse by designing a two-frequency waveform aimed at nucleating and controlling the collapse of bubble clouds on kidney stones (Ikeda et al., 2006; Yoshizawa et al., 2009; Matsumoto and Yoshizawa, 2005; Brujan and Matsumoto, 2012). This method, sometimes called cavitation control lithotripsy (CCL), employs an initial focused ultrasound pulse with very high frequency (usually 1 to 5 MHz) to create a localized bubble cloud. A second pulse at or around 545 khz trails the high frequency pulse, and forces the bubble cloud to collapse near the stone surface, delivering high pressures to erode the stone in the process. This approach exploits cloud cavitation to erode the stone. While the pressure magnitudes used for CCL are marginally higher than those of BWL — Yoshizawa et al. (2009) report peak positive focal pressure of 16 MPa — it is an efficient method to deliver very high localized pressure on the stone surface, which can reach $O(1\,\text{GPa})$ as the bubble cloud collapses (Shimada, Kobayashi, and Matsumoto, 1999).

By contrast to SWL and CCL, the resonating stress waves associated with the ultrasound wavelength appear to be sufficient to produce small enough fragments in burst-wave lithotripsy without relying on cavitation-induced damage (Maxwell, Cunitz, et al., 2015; Maxwell, MacConaghy, et al., 2020; Chen, Samson, et al., 2020; Raskolnikov, Bailey, and Harper, 2022). In developing treatment protocols, it is therefore generally sought to avoid cavitation through pausing treatment to allow bubbles to dissolve, or to utilize interleaved pulses of lower-amplitude ultrasound to induce bubble coalescence and dissolution (Duryea, Roberts, et al., 2014; Duryea, Cain, et al., 2015; Duryea, Tamaddoni, et al., 2015; Tamaddoni, Roberts, Duryea, et al., 2016; Tamaddoni, Roberts, and Hall, 2019). Apart from minimizing the risk of injury, minimizing cavitation is important in BWL because the short inter-pulse times do not allow the full dissolution of bubbles from pulse to pulse. The continual presence of bubbles on the proximal surface of the stone can then lead to a substantial shielding of the stone from the incident acoustic energy (Maeda, Maxwell, et al., 2018).

To better understand the mechanisms of stone comminution during BWL, a tool capable of simulating the response to BWL of an arbitrary stone, while modeling the dynamics of bubbles that may shield the stone during treatment, would be valuable.

1.7 Fluid-Solid Solver

To accurately simulate ultrasound–bubble cloud interactions, and cavitation of bubbles which may be present during treatment, a high-order accurate multi-phase compressible flow solver is required. To this end, we use an interface capturing method with the so-called 5-equation multiphase flow model of Kapila et al. (2001), solved with a WENO shock-capturing finite volume scheme. This is paired with an HLL-type Riemann solver and total-variation diminishing time-stepping. The solver used, now open source, is the Multi-component Flow Code (MFC). Also implemented is a model to propagate spherical acoustic waves, as generated by the transducer during BWL. Previous numerical studies with this solver have examined the energy shielding effects of bubbles by using a subgrid bubble model and calculating the scattering of acoustic energy (Maeda, Maxwell, et al., 2018; Maeda and Colonius, 2018; Maeda and Colonius, 2019). However, these did not include a solid mechanics model or any way to determine the elastic response of the stone to the ultrasound.

Various approaches exist to pair a solid-mechanical model with a fluid dynamics solver. A difficulty in this case is that modeling the response of a solid generally requires tracking material deformation. However, the Eulerian framework used in MFC and other structured-grid CFD solvers does not track strains, and thus the stresses in the solids cannot be calculated using typical stress-strain relations. To address this problem and maintain a single-framework, monolithic solver, we adopt a hypoelastic material model following Rodriguez and Johnsen (2019). This approach is capable of modeling solid mechanics in a fully Eulerian framework by taking appropriate derivatives of constitutive stress-strain relations. To determine elastic stresses in this framework, we only require strain derivatives, which are straightforward to calculate in an Eulerian framework where velocity is tracked at each point in the domain. Thus, such a hypoelastic solver can naturally be implemented in a CFD solver such as MFC, without the need to switch the frame of reference used. Details on the hypoelastic approach and its implementation are discussed in Chapter 4, in section 4.1.1. There, we also detail the addition of a continuum damage model following Cao et al. (2019). This model makes use of the stresses obtained in the stone via the hypoelastic model to calculate a damage field, which tracks the accumulation of high stresses over time in the stone. Thus, the full interaction of ultrasound waves with gas bubbles and a submerged stone can be modeled to better understand mechanisms key to improving the efficacy of BWL.

1.8   Contributions and Outline

In this book, we first present the development of a data assimilation framework for the characterization of soft, viscoelastic materials via observation of bubble collapse (DA-IMR). This includes the implementation of a hybrid DA method which is particularly effective for parameter estimation in this context. We show its successful application to data from various experiments using different constitutive models, demonstrating the accuracy and versatility of material property estimation with DA-IMR. Beyond parameter estimation, we gain insight into the limitations of the physical models used when violent bubble collapse occurs, particularly with high energy laser-induced cavitation. We then present the implementation of a comprehensive numerical framework for high-fidelity direct numerical simulations of acoustic wave–bubble–stone interactions. This framework is used to investigate bubble dynamics and stone damage in burst-wave lithotripsy. The reduction of stress and damage in stones during BWL when bubbles are present is quantified, and strategies to mitigate bubble shielding and improve treatment efficacy by modulating the frequency of individual transducer elements are proposed.

Chapter 2 details the development of the DA-IMR framework. A reduced-order spherical bubble dynamics model, and various ensemble-based data assimilation methods used in conjunction with this model are presented. Surrogate bubble radius time history data is generated via numerical simulations of inertial bubble collapse, and used to compare the performance of each data assimilation method. We conclude on the most accurate and efficient methods in the context of soft material characterization, and validate the DA-IMR framework with these DA methods.

In Chapter 3, this framework is tested on experimental cavitation data in three subsequent studies. We begin by applying DA-IMR on the laser-induced cavitation data of Estrada et al. (2018), which serves as a benchmark for our method with experimental data. Then, we apply DA-IMR to results with laser-induced cavitation (LIC) in gels with seed particles, which expands the stretch-ratio regime where DA-IMR can be used. The DA methods not only estimate the mechanical properties of the gel, but show the limitations of the physical model used for violently cavitating bubbles. Finally, the method is tested with ultrasound-induced cavitation data, which we show avoids some of the issues observed with bubble nucleation physics in LIC.

Next, in Chapter 4, we shift to direct numerical simulations of acoustic wave–bubble–stone interactions. This chapter describes the developed framework in detail, including all the implementations to make BWL simulations possible. We begin by describing the multiphase flow solver used for the interactions of acoustic waves with liquids and gas bubbles. Then, we present the hypoelastic model implemented into this solver to simulate the elastic response of solids, and validate it against analytical and experimental results. Additionally, we present the addition of a continuum damage model, the calibration of a virtual transducer array to mimic experimental conditions with a much smaller simulated acoustic source, and the acceleration of our code using GPUs to enable simulations at high enough resolutions to capture all scales required in BWL simulations. Example simulations are performed to study the ultrasound frequency–dependence of maximum stress in stones of various shapes and sizes.

In Chapter 5, this numerical framework is used to study bubble dynamics and damage in the stone during BWL. We begin by quantifying the damage mitigation in the stone due to bubble shielding, showing the significant impact that bubble clouds can have on treatment efficacy. We then study the dynamics of characteristic bubble clouds when exposed to spherically focused ultrasound acoustic fields generated by a virtual BWL transducer array. We show that modulating the ultrasound frequency of individual transducer elements by introducing a secondary low-frequency wave can cause bubbles to collapse ahead of the stone. Applying this to full BWL simulations with bubble clouds present in front of a model kidney stone, we explore strategies to maximize the efficiency and efficacy of single BWL pulses in the presence of bubbles.

Finally, in Chapter 6, we summarize the conclusions of this book and provide guidance on potential future efforts to expand on the work presented.

Chapter 2

DATA ASSIMILATION FOR INERTIAL MICROCAVITATION RHEOMETRY

Part of this chapter is adapted from sections 1-4 of Spratt, Rodriguez, Schmidmayer, et al. (2021). We begin by describing the material–bubble-dynamic model used for inertial microcavitation rheometry throughout Chapter 2 and Chapter 3, with slight modifications to the viscoelastic material model where needed. Then, we present three data assimilation methods and details of their implementation for our problem. Finally, we compare each method using synthetic data, generated by adding random noise to results from our spherical bubble dynamics model. Thus, we can gauge the relative performance of each method in the absence of modeling uncertainties. We show that two of the tested methods can predict material parameters accurately, and are thus chosen as candidates to use with experimental data in Chapter 3.

2.1   Bubble Dynamics Model

A physical model for the dynamics of collapsing bubbles is required to character-ize the viscoelastic properties of surrounding materials. Many spherical bubble dynamics models exist. Of particular relevance here are those for cavitation in soft materials (Gaudron, Warnez, and Johnsen, 2015; Yang and Church, 2005) and specific numerical methods for solving them (Warnez and Johnsen, 2015; Barajas and Johnsen, 2017). We use the model of Estrada et al. (2018), which adopts ap-proximations validated in previous spherical-bubble models (Prosperetti and Lezzi, 1986; Prosperetti, Crum, and Commander, 1988; Akhatov et al., 2001; Epstein and Keller, 1972; Keller and Miksis, 1980; Preston, Colonius, and Brennen, 2007). Key assumptions of this model are that the motion of the bubble and its contents are spherically symmetric, the bubble pressure is spatially uniform (homobaricity), the temperature of the surrounding material is constant, and that there is no mass trans-fer of the non-condensible gas across the bubble wall. While the effects of vapor mass transfer may be negligible in the regime of interest to this chapter (Barajas and Johnsen, 2017), we opt to include these in the present model formulation as they are important in other regimes where this method may be applied (Preston, Colonius, and Brennen, 2007).

The Keller–Miksis equation models the radius evolution (Keller and Miksis, 1980),

$$\left(1 - \frac{\dot{R}}{c}\right) R\ddot{R} + \frac{3}{2}\left(1 - \frac{\dot{R}}{3c}\right)\dot{R}^2 = \frac{1}{\rho}\left(1 + \frac{\dot{R}}{c}\right)\left(p_b - \frac{2s}{R} + S - p_\infty\right) \\ + \frac{1}{\rho}\frac{R}{c}\overline{\left(p_b - \frac{2s}{R} + S\right)},$$
(2.1)

where R is the bubble radius, c the material speed of sound, ρ the material density, p_b the bubble internal pressure, s the bubble-wall surface tension, S the stress integral (see (2.7)), and p_∞ the far-field pressure. The Rayleigh-Plesset equation (Rayleigh, 1917; Plesset, 1948; Plesset, 1949; Plesset and Prosperetti, 1977) is often used for spherical bubble dynamics. However, it assumes incompressibility of the surrounding material, whereas the Keller-Miksis equation relaxes the incompressibility assumption near the bubble wall, taking into account the speed of sound of the surrounding material. These effects prove to be important in the context of IMR, though values for the Mach number are not expected to exceed $\dot{R}/c \approx 0.3$ (Akhatov et al., 2001), and thus the Keller-Miksis equation is used.

Under the model assumptions, no mass or energy conservation equations are needed outside the bubble. Furthermore, the conservation of momentum simplifies to an ordinary differential equation for the bubble pressure

$$\dot{p}_b = \frac{3}{R}\left[-\kappa p_b \dot{R} + (\kappa - 1)K(T(R))\left.\frac{\partial T}{\partial r}\right|_{r=R} + \kappa p_b \frac{C_{p,v}}{C_p(k_v(R))}\frac{D}{1 - k_v(R)}\left.\frac{\partial k_v}{\partial r}\right|_{r=R}\right],$$
(2.2)

where κ is the specific heat ratio of vapor, K the thermal conductivity, T the gas temperature, C_p the specific heat, D the binary diffusion coefficient, and k_v the vapor mass fraction. Subscripts g, v, and m refer to gas, vapor, and mixture properties, respectively. Applying conservation of energy in the bubble interior yields an equation for the bubble temperature:

$$\rho_m C_p\left(\frac{\partial T}{\partial t} + v_m \frac{\partial T}{\partial r}\right) = \dot{p}_b + \frac{1}{r^2}\frac{\partial}{\partial r}\left(r^2 K \frac{\partial T}{\partial r}\right) + \rho_m(C_{p,v} - C_{p,g})D\frac{\partial k_v}{\partial r}\frac{\partial T}{\partial r},$$
(2.3)

where v_m is the radial mixture velocity and ρ_m the mixture density. The boundary condition $T(R) = T_\infty$ follows from the model assumptions. The radial mixture velocities are computed as

$$v_m(r, t) = \frac{1}{\kappa p_b}\left[(\kappa - 1)K\frac{\partial T}{\partial R} - \frac{1}{3}r\dot{p}_b\right] + \frac{C_{p,v} - C_{p,g}}{C_{p,m}}D\frac{\partial k}{\partial r},$$
(2.4)

with associated kinematic boundary condition

$$v_m(R) = \dot{R} - \frac{D}{1 - k_v(R)} \left.\frac{\partial k_v}{\partial r}\right|_{r=R}. \tag{2.5}$$

Fick's law describes the mass diffusion process in the bubble. Casting the conservation of mass inside the bubble in terms of the mixture density, the vapor mass fraction inside the bubble is

$$\frac{\partial k_v}{\partial t} + v_m\frac{\partial k_v}{\partial r} = \frac{1}{\rho_m}\frac{1}{r^2}\frac{\partial}{\partial r}\left(r^2\rho_m D\frac{\partial k_v}{\partial r}\right). \tag{2.6}$$

Under the assumption of equilibrium phase change at the bubble wall, the associated boundary condition at the wall is $p_{v,sat}(T(R)) = \mathcal{R}_v k(R)\rho_m(R)T(R)$, where $p_{v,sat}$ is the saturation pressure of the vapor and $\mathcal{R}_v$ is the gas constant of the vapor.

Equations (2.1), (2.2), (2.3), and (2.6) form a system of equations, which is evolved in time with an implicit Runge–Kutta algorithm that uses the trapezoidal rule and backwards differentiation at each step (TR–BDF2) (Hosea and Shampine, 1996; Shampine and Reichelt, 1997; Shampine, Reichelt, and Kierzenka, 1999). The partial differential equations for temperature and vapor mass fraction are discretized in space via a uniform grid and computed using second-order-accurate central finite differences. Estrada et al. (2018) showed that the finite-deformation neo-Hookean Kelvin–Voigt model can adequately represent the material response at high strain rates in problems of interest, and that a more complicated model did not improve the goodness of fit of simulated radius curves with experimental data. In this framework, the material is modeled with a parallel spring (neo-Hookean elastic response with shear modulus G) and dashpot (linear viscous response with viscosity μ). The stress integral in (2.1) is

$$S = -\frac{G}{2}\left[5 - \left(\frac{R_0}{R}\right)^4 - 4\frac{R_0}{R}\right] - \frac{4\mu\dot{R}}{R}, \tag{2.7}$$

where R_0 is the equilibrium bubble radius (Gaudron, Warnez, and Johnsen, 2015). Details on the derivation of this spherical bubble dynamics model, and comparison with the model of Preston, Colonius, and Brennen (2007), are shown in Appendix A.

2.2 Data Assimilation Methods

Two difficulties that drive the choice of data assimilation method are the nonlinearity of the dynamics and large state vector required to discretize the partial differential equations adequately. The former rules out the standard linearized Kalman filter (EKF) (Kalman, 1960) and the latter renders its direct nonlinear extensions (e.g., the unscented Kalman filter, UKF) computationally prohibitive. Instead, ensemble-based methods (Evensen, 1994) are considered. They combine computational efficiency with nonlinear dynamics by approximating the state covariance via statistics of a finite (and typically small) ensemble. We consider three specific ensemble methods: an ensemble Kalman filter (EnKF), an iterative ensemble Kalman smoother (IEnKS), and a hybrid ensemble-based 4D–Var method.

The discretized equations of section 2.1 are rewritten as a nonlinear operator F, and we define the linear observation function H that maps the state x to measurement space. This yields the discrete-time dynamical system

$$x_{k+1} = F(x_k) + \eta_k, \tag{2.8}$$

$$y_k = H(x_k) + v_k, \tag{2.9}$$

where

$$x_k \in \mathbb{R}^d , \; y_k \in \mathbb{R}^n,$$

$$\eta_k \sim \mathcal{N}(0, \Sigma) , \; v_k \sim \mathcal{N}(0, \Gamma),$$

$$F : \mathbb{R}^d \to \mathbb{R}^d , \; H : \mathbb{R}^d \to \mathbb{R}^n.$$

x_k is the d-dimensional state at time k comprised of all the dependent variables plus the unknown parameters

$$x = \{R, \dot{R}, p_b, S, \mathbf{T}, \mathbf{C}, \log(\text{Ca}), \log(\text{Re})\}, \tag{2.10}$$

which are the bubble-wall radius, velocity, bubble pressure, stress integral, the discretized temperature and vapor concentration fields inside the bubble, and the log-Cauchy and log-Reynolds numbers, respectively. The Cauchy and Reynolds numbers are here defined as

$$\text{Ca} \equiv \frac{p_\infty}{G} \quad , \quad \text{Re} \equiv \frac{\sqrt{\rho p_\infty} R_{\max}}{\mu}. \tag{2.11}$$

These quantities appear in the nondimensionalized model equations of section 2.1 for the surrounding material and the shear modulus G and viscosity μ, our material properties of interest, can be computed via (2.11). When the state vector is stepped forward in time with equation (2.8), the forecast operator F simply maps $\log(\text{Ca})$ and $\log(\text{Re})$ to themselves, as they are constant in the physical model. Appending them to the state vector enables the estimation of these parameters within the data assimilation framework in a straightforward way. We note that the logarithm of these parameters is taken to avoid negative (and thus non-physical) values of the Cauchy and Reynolds numbers (Ca and Re) occurring during the analysis step of the assimilation algorithms (described in sections 2.2.1, 2.2.2, and 2.2.4).

The variable y_k is the n-dimensional observation (data) at time k. η_k is the unknown process noise (or model error) added to $H(x_k)$ to retrieve y_k. It is assumed to be Gaussian with zero mean and standard deviation Σ. Similarly, v_k is the assumed Gaussian measurement noise added to $F(x_k)$ to obtain x_{k+1}, with zero mean and unknown standard deviation Γ. Throughout this study, the only available measurement is the bubble radius. This means that y_k is comprised of a single element: the bubble radius. The observation operator H is then the linear map from the state vector to its first element R. In the following, the linear operator H is sometimes represented as the matrix H for clarity ($Hx = H(x)$).

The following methods estimate the full state vector x (including parameters of interest $\log(Ca)$ and $\log(Re)$) based on observations of y. The EnKF and IEnKS are online (or quasi-online) methods—they optimize the value of x at each time through the simulation. The IEnKS is deemed quasi-online because it uses data from a few future times as well. The estimation trails the simulation time by a fixed number of time steps called the lag. Alternatively, the En4D–Var is an offline method, which only optimizes the initial condition for x, taking into account data from the entire time domain.

2.2.1   The Ensemble Kalman Filter

The ensemble Kalman filter (Evensen, 1994) represents the probability density function (PDF) for the state of the dynamics through the statistics of an ensemble of q state vectors. It does not require an adjoint, or deriving a tangent linear operator to the physical model (Evensen, 2003; Evensen, 2009a; Evensen, 2009b). Starting with suitably randomized initial conditions, each ensemble member is propagated through the physical model, and the predictions are then corrected using the ensemble statistics. The ensemble is initialized with a guess for the initial condition x_0 as the mean, and a given covariance corresponding to the expected error covariance. In practice, each ensemble member is independently sampled from a normal distribution with mean x_0 and the assumed covariance matrix. Several initialization strategies exist depending on the system and its dynamics. In the present case, the nonlinear dynamics render a systematic approach difficult. Instead, the ensemble is initialized with a covariance corresponding to our best estimate based on simulations, and adjusted through trial and error to optimize results. An ensemble size of $q = 48$ is used for the tests. This has been shown to give accurate results while keeping computational costs modest. Indeed, in testing with the current framework, larger ensemble sizes led to negligible improvement in parameter estimation but significantly increased the cost of tested DA methods. At any given time, the estimated value for the state vector is then taken to be the ensemble average.

$$\overline{x}_k = \frac{1}{q} \sum_{j=1}^{q} x_k^{(j)}. \tag{2.12}$$

The filter is broken down into a forecast and an analysis step. In the forecast step, the physical model is used to step the state forward in time with (2.8). Each representation of the state vector $x_k^{(j)}$ in the ensemble at time k is propagated through F with $\hat{x}_{k+1}^{(j)} = F(x_k^{(j)})$. Next, in the analysis step, if an experimental measurement y_{k+1} is available at the current time step $k+1$, then it is used to correct the forecast. As described in the dynamical system equations, each ensemble member is mapped to measurement space $H(x_{k+1})$. The analysis proceeds by minimizing a cost function involving the difference between $H(x)$ and the data point y, while accounting for measurement noise and model error. This cost function is similar to that of the Kalman filter and is given by

$$J(x) = \frac{1}{2}\|y_k - H(x)\|_R^2 + \frac{1}{2}\|x - \hat{x}_k\|_{C_k}^2 \tag{2.13}$$

$$= \frac{1}{2}[y_k - H(x)]^T R^{-1}[y_k - H(x)] + \frac{1}{2}[x - \hat{x}_k]^T C_k^{-1}[x - \hat{x}_k], \tag{2.14}$$

where R is the measurement noise covariance matrix, which is an input to the algorithm, and C_k is the ensemble covariance at time step k. A key difference with the Kalman filter is that this covariance is calculated empirically in this case. It is defined as

$$C_k = A_k(A_k)^T, \tag{2.15}$$

where A_k is the state perturbation matrix

$$A_k = \frac{1}{\sqrt{q-1}} \left[x_k^{(1)} - \overline{x}_k, \ \dots, \ x_k^{(q)} - \overline{x}_k \right]. \tag{2.16}$$

In fact, the minimization does not make use of the covariance matrix directly, but instead uses the state perturbation matrix and scaled output perturbation matrix HA_k defined as

$$HA_k = \frac{1}{\sqrt{q-1}} \left[y_k^{(1)} - \overline{y}_k, \ \dots, \ y_k^{(q)} - \overline{y}_k \right]. \tag{2.17}$$

The optimization is carried out by finding the minimizer x_k satisfying

$$x_k = \hat{x}_k + A_k \cdot w_k, \tag{2.18}$$

with w_k a correction coefficient. This restricts the solution to the subspace spanned by the scaled perturbation matrix around the prior estimate $\hat{x}_k$. The optimization can be restated as

$$w_k = \underset{w \in \mathbb{R}^q}{\operatorname{argmin}} J(w), \tag{2.19}$$

where

$$J(w) = \frac{1}{2}\|w\|^2 + \frac{1}{2}\|y_k - H(\hat{x}_k) - HA_k(w)\|_R^2. \tag{2.20}$$

The solution is unique, and using the Woodbury matrix identity to write the inversion in measurement space, can be written as

$$w_k = (HA_k)^{\mathrm{T}}[R + (HA_k)(HA_k)^{\mathrm{T}}]^{-1}(y_k - H(\hat{x}_k)). \tag{2.21}$$

Performing this inversion in the measurement space is in most cases more computationally efficient. Here this is clear, as the measurement space is comprised of only one variable (the bubble radius). Once the minimizer is found and the analysis step complete, covariance inflation is applied to the ensemble to correct for the (typical) underestimation of the variance with finite (typically small) ensembles (see section 2.2.3 for details on covariance inflation). Finally, the forecast step can be repeated. Figure 2.1 shows a flow chart of the EnKF method for a visual representation of the steps described above. This flow chart will remain the same for the IEnKS method described in section 2.2.2, with changes to the analysis step only.

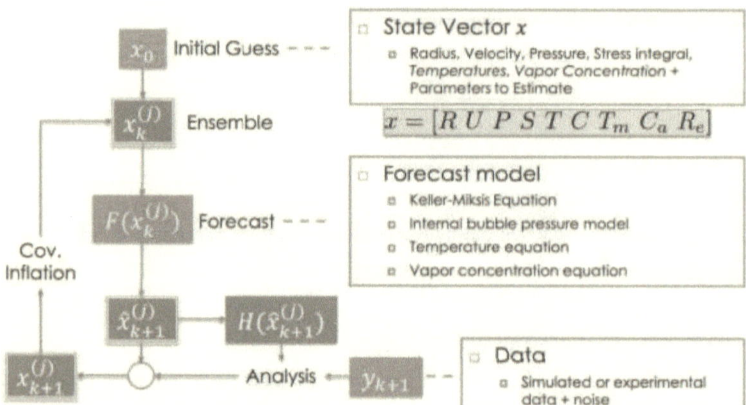

$$x = \begin{bmatrix} R\ U\ P\ S\ T\ C\ T_m\ C_a\ R_e \end{bmatrix}$$

Figure 2.1: Flowchart of the Ensemble Kalman Filter as applied in the framework of inertial microcavitation rheometry.

2.2.2 The Iterative Ensemble Kalman Smoother

Minimizing deviation from data not just at the current estimation step, but at future times, can help to smooth out estimation and focus on longer-term trends. The IEnKS uses information from one or multiple future time steps in its assimilation, and can thus be an effective tool. While the ensemble initialization and forecast step are the same as that of the EnKF, the difference in the analysis step is twofold. First, the cost function is modified to minimize difference with data at a single or multiple future times (Evensen and Leeuwen, 2000). The assimilation thus trails the simulation by a number of time steps (called the lag). Second, it is no longer minimized analytically but iteratively using a Gauss–Newton algorithm.

The IEnKS method used here is from Bocquet and Sakov (2013a) and Sakov, Oliver, and Bertino (2012). Bocquet and Sakov (2013b) have shown it to be effective for state and parameter estimation problems with highly nonlinear dynamics. Its cost function can take two forms referred to as 'single data assimilation' (SDA) or 'multiple data assimilation' (MDA) (Bocquet and Sakov, 2013a). The IEnKS–SDA cost function penalizes difference with measurements at a single time step $k + L$, where L corresponds to the lag of the smoother. It is given by

$$J(x) = \frac{1}{2}\|y_{k+L} - H \circ F_{k \to (k+L)}(x)\|_R^2 + \frac{1}{2}\|x - \hat{x}_k\|_{C_k}^2, \tag{2.22}$$

where $H \circ F$ denoted the composition of H and F.

On the other hand, the IEnKS–MDA cost function minimizes this difference over a data assimilation window (DAW) from $k + 1$ to $k + L$, and is expressed as

$$J(x) = \frac{1}{2} \sum_{i=1}^{L} \beta_i \| y_{k+i} - H \circ F_{k \to (k+i)}(x) \|_R^2 + \frac{1}{2} \| x - \hat{x}_k \|_{C_k}^2, \tag{2.23}$$

where β_i are weights attributed to given time steps with $\sum \beta_i = 1$. Again, a solution of the form $x = \hat{x}_k + A_k \cdot w$ is sought, but a Gauss–Newton method is used (Bocquet and Sakov, 2013a). The minimizer w is found by iterating following

$$w_{(i+1)} = w_{(i)} - \mathcal{H}_{(i)}^{-1} \Delta J_{(i)}(w_{(i)}), \tag{2.24}$$

where i is the iteration number, and $\mathcal{H}$ is the approximate Hessian

$$\mathcal{H}_{(j)} = (q - 1)I + HA_{(j)}^{\mathrm{T}} R^{-1} HA_{(j)}, \tag{2.25}$$

where I is the $q \times q$ identity matrix. The gradient is given by

$$\Delta J_{(j)} = -HA_{(j)}^{\mathrm{T}} R^{-1} [y_{k+L} - H \circ F_{k+L \leftarrow k}(x_k)] + (q - 1)w_{(j)}. \tag{2.26}$$

For the smoother, HA is more complicated than it is for the filter as it involves differences with measurements at future time steps. This quantity is akin to a tangent linear operator from ensemble to measurement space and has to be estimated. Following Bocquet and Sakov (2013a), a finite-difference estimate is used:

$$HA_{(j)} \approx \frac{1}{\alpha} H \circ F_{k+L \leftarrow k}(x_k^{(j)} \mathbf{1}^{\mathrm{T}} + \alpha A_k) \left(I - \frac{\mathbf{1} \cdot \mathbf{1}^{\mathrm{T}}}{q} \right), \tag{2.27}$$

with scaling factor $\alpha \ll 1$ and $\mathbf{1} = (1 \; \cdots \; 1)^{\mathrm{T}}$ a vector of length q. The iteration is repeated until a threshold $w_{(i+1)} - w_{(i)} < \epsilon$, or a fixed number of iterations is reached. Once the optimal value w_{opt} is obtained, $x_{\mathrm{opt}} = \hat{x} + A_k \cdot w_{\mathrm{opt}}$ is calculated and a new ensemble E_k is sampled at time step k with

$$E_k = x_{\mathrm{opt}} \mathbf{1}^{\mathrm{T}} + \sqrt{q - 1} A_k \mathcal{H}_{\mathrm{opt}}^{-1/2} I. \tag{2.28}$$

This completes the analysis step. When using the MDA variant, the Hessian and gradient of J are found with

$$\mathcal{H}_{(j)} = (q - 1)I + \sum_{i=1}^{L} HA_i^{\mathrm{T}} \beta_i R^{-1} HA_i \tag{2.29}$$

$$\Delta J_{(j)} = -\sum_{i=1}^{L} HA_i^{\mathrm{T}} \beta_i R^{-1} [y_{k+i} - H \circ F_{k+i \leftarrow k}(x_i)] + (q - 1)w_{(j)}. \tag{2.30}$$

2.2.3 Covariance Inflation

While the EnKF and IEnKS may converge, ensemble methods are subject to intrinsic sampling error (Bocquet, 2011; Luo and Hoteit, 2011). This sampling error results from the finite ensemble size q used to represent the statistics of a system of often much higher dimension. As Leeuwen (1999) explains, the EnKF tends to underestimate error variances, particularly for small ensemble sizes. There exist different ways to address this sampling error, but a simple approach is covariance inflation (Whitaker and Hamill, 2012), where we correct

$$x^{(j)} = \overline{x} + \alpha(x^{(j)} - \overline{x}) + \lambda^{(j)}. \tag{2.31}$$

Here, $\overline{x}$ denotes the ensemble average, as defined in (2.12), after the analysis step. Parameters α and λ correspond to multiplicative and additive inflation parameters, respectively.

There exist many schemes for multiplicative inflation, the most simple of which is picking a scalar α (usually $1.005 \le \alpha \le 1.05$). This can work well but requires extensive tuning to optimize the value for each run or data set. Instead, Whitaker and Hamill (2012) propose a scheme they call 'Relaxation to Prior Spread' (RTPS). Here, the value for α is found at each time step using

$$\alpha_i = 1 + \theta \left(\frac{\sigma_i^b - \sigma_i^a}{\sigma_i^a} \right), \tag{2.32}$$

where σ_i^a and σ_i^b are the prior and posterior ensemble standard deviation for the ith element of the state vector (α is a vector here), and θ is a scalar (usually $0.5 \le \theta \le 0.95$). As this expression for α shows, this scheme inflates the covariance more in regions where the analysis led to a large correction. Whitaker and Hamill (2012) test this method and compare it to other approaches, showing that it performs well by adequately preventing underestimation of error variances and better converging on accurate estimates. We tested this covariance inflation scheme, and similarly found that it performs as well or better than a simple scalar α for data assimilation in our context. After some tuning across data sets, this RTPS model was used with $\theta = 0.7$. Additive covariance inflation was not found to significantly affect results and introduced some stability issues with larger magnitudes of λ. Therefore, $\lambda = 0$ is used.

2.2.4 A hybrid Ensemble-based 4D–var Method

As with ensemble Kalman methods, ensembles can be used with 4D-Var to estimate covariance empirically, thus reducing computational cost (Gustafsson and Bojarova, 2014; Liu, Xiao, and Wang, 2008; Caya, Sun, and Snyder, 2005; Trémolet, 2007). The present method (En4D-Var) is a fully offline extension of the IEnKS–MDA method. Again, the ensemble is initialized in the same way as EnKF, but the cost function is here

$$J(\boldsymbol{x}) = \frac{1}{2} \sum_k \beta_k \|\boldsymbol{y}_k - H \circ F_{k \leftarrow 0}(\boldsymbol{x})\|_{\boldsymbol{R}}^2 + \frac{1}{2} \|\boldsymbol{x} - \hat{\boldsymbol{x}}_k\|_{C_0}^2. \qquad (2.33)$$

The difference with the IEnKS–MDA cost function is the data assimilation window size. Rather than minimizing over a few time steps forward and then stepping through time, the minimization is done over the entire time domain and only the initial state vector is corrected. Each new iteration is initialized with the corrected initial state (including parameters to estimate). The same minimization procedure as described in section 2.2.2 is used. When the minimization has converged, a final simulation is run with the forecast model only. In cases where only a few iterations are necessary, this method reduces computational cost as compared to the IEnKS-MDA. The time dimension is still fully included, but each point in time is only assimilated once per iteration. Furthermore, this retains the advantage of ensemble methods. As opposed to classical 4D–Var, there is no need to linearize the state function and find the tangent linear adjoint operator. This novel adaptation of the IEnKS method is well suited to the present problems given that our interest is the estimation of material properties which are, at the outset, assumed to be constant.

Whether online or offline, the structure of all the presented methods provides a significant computational advantage, as compared to the least squares fitting used by Estrada et al. (2018). For simple least squares fitting, a large array of cases with varying G and μ need to be simulated. This requires $N_G \times N_\mu$ simulations, where N_G and N_μ are the number of shear moduli and viscosities to test. To obtain precise estimates (to several decimal points), these must be large. On the other hand, only an initial guess for these parameters is needed in the data assimilation methods presented. Here, the cost will increase with increasing ensemble size q (at different rates depending on the method), as each ensemble member is stepped through time independently. However, a small value of $q = 48$ is shown to suffice for the present estimation, resulting in low computational time while providing good parameter estimates for most methods, as shown in sections 2.3.1 and 3.1.2. This advantage becomes even more apparent if the method is scaled to estimate more parameters. For least squares fitting, the number of simulations to run would be multiplied by the number of values N to test for each new parameter, quickly becoming unfeasible for precise estimation. On the other hand, these data assimilation methods would incur a minimal added computational cost associated with the added state variable. In this chapter, we only estimate two parameters given our constitutive material model, namely G and μ, found through estimation of $\log(\mathrm{Ca})$ and $\log(\mathrm{Re})$. However, we will see examples in Chapter 3 where we estimate up to 4 material parameters given a different viscoelastic model. There, the number of simulations to run with simple least-squares fitting would be $\prod_{i=1}^{4} N_i$, with each N_i the number of guesses for each parameter i to estimate. In contrast, the added computational cost for any of the data assimilation methods presented is negligible, as we are simply adding a single additional element to the state vector which has $O(100)$ elements.

2.3 Framework Validation with Surrogate Data

2.3.1 Parameter Estimation Results

Synthetic data where the true shear modulus and viscosity are known is generated from the model (section 2.1) and used to test the data assimilation methods in a setting where there is no modeling error. Bubble radius time-history data from the simulation is sampled every 3.7 µs to match the 270,000 frames per second image capture rate in available experiments. Random Gaussian noise is added to these samples to mimic experimental data. The standard deviation of this noise is set at $\sigma = 0.02$, which is greater than the estimated noise in the experiments. Two polyacrylamide gels were examined with nominal values of shear modulus and viscosity determined by Estrada et al. (2018). For the stiff gel: $G_{\mathrm{stiff}} = 7.69\,\mathrm{kPa}$, $\mu_{\mathrm{stiff}} = 0.101\,\mathrm{Pa\,s}$, and for the soft gel: $G_{\mathrm{soft}} = 2.12\,\mathrm{kPa}$, $\mu_{\mathrm{soft}} = 0.118\,\mathrm{Pa\,s}$. Since similar estimation accuracy was achieved in both cases, we report results for the stiff gel only. The other material properties used are taken from Estrada et al. (2018) and given in table 2.1. No uncertainty is added to these parameters in the present study to match their conditions and focus on estimating G and μ.

An example simulated radius curve and sampled surrogate measurements (with noise added) with these parameters is shown in figure 2.2a, plotted against non-dimensional time

$$t^* = \frac{t}{R_{\max}} \sqrt{\frac{p_\infty}{\rho}}. \tag{2.34}$$

Parameter	Value	Parameter	Value
ρ	$1060\,\mathrm{kg/m^3}$	c	$1430\,\mathrm{m/s}$
p_∞	$101.3\,\mathrm{kPa}$	s	$5.6 \times 10^{-2}\,\mathrm{N/m}$
D	$24.2 \times 10^{-6}\,\mathrm{m^2/s}$	κ	1.4
$C_{p,g}$	$1.62\,\mathrm{kJ/kg\,K}$	$C_{p,v}$	$1.00\,\mathrm{kJ/kg\,K}$
A	$5.3 \times 10^{-5}\,\mathrm{W/m\,K^2}$	B	$1.17 \times 10^{-2}\,\mathrm{W/m\,K^2}$
p_{ref}	$1.17 \times 10^8\,\mathrm{kPa}$	T_{ref}	$5200\,\mathrm{K}$
T_∞	$298.15\,\mathrm{K}$		

Table 2.1: Model parameters as they follow from Estrada et al. (2018). Table reprinted from Spratt, Rodriguez, Schmidmayer, et al. (2021) with permission from Elsevier, © 2021.

With the simulated data, the evolution over time of all variables in the state vector is known. For example, bubble-wall velocity, bubble pressure and stress integral are plotted in figure 2.2b.

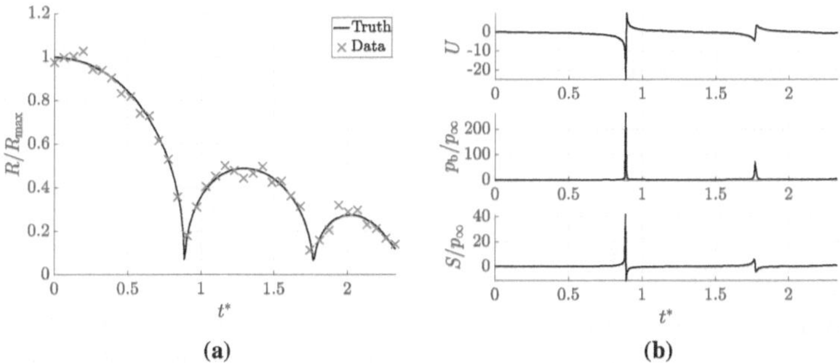

(a)   (b)

Figure 2.2: Simulated bubble radius and noisy sampled data used to test data assimilation methods (a), alongside simulated bubble-wall velocity, normalized bubble pressure and stress integral (b), plotted over non-dimensional time t^*. Reprinted from Spratt, Rodriguez, Schmidmayer, et al. (2021) with permission from Elsevier, © 2021.

A set of initial guesses for the shear modulus and viscosity, ranging from 10% to 100% error from the true values, were used to test each method. Table 2.2 summarizes results for a subset of these cases, representing 10, 50, and 100% initial error in G and μ. In each case, ensembles were initialized as Gaussian with these erroneous material properties as the mean, and standard deviation increasing with increased error. That is, the spread of the initial ensemble was made wider for cases with more error, to account for the increased uncertainty in the initial guess.

To match the experimental data (as used in section 3.1 in Chapter 3), simulated data is limited to the first three peaks of the bubble collapse. This corresponds to approximately 35 points given the initial conditions and frame rate. Estrada et al. (2018) found that limiting the data to this region led to better parameter estimation. Similarly, we find that the model fails to fit the radius measurements after this time. Reasons for this reduced accuracy at later times are discussed in results with experimental data in Chapter 3.

Method	G estimate (%error) [kPa]	μ estimate (%error) [Pa s]	Run time [s]
Guess 1	8.50 (+10%)	0.09 (-10%)	–
EnKF	7.234 (5.93%)	0.098 (2.61%)	428
IEnKS–SDA (lag 1)	7.364 (4.24%)	0.110 (8.58%)	852
IEnKS–MDA (lag 3)	6.682 (13.11%)	0.100 (0.92%)	8076
En4D–Var	7.150 (7.03%)	0.099 (1.80%)	679
Guess 2	3.80 (-50%)	0.05 (-50%)	–
EnKF	3.988 (48.1%)	0.057 (43.1%)	375
IEnKS–SDA (lag 1)	4.203 (45.4%)	0.080 (20.9%)	904
IEnKS–MDA (lag 3)	7.390 (3.90%)	0.086 (15.2%)	9755
En4D–Var	7.396 (3.82%)	0.100 (0.52%)	690
Guess 3	15.0 (+100%)	0.20 (+100%)	–
EnKF	13.649 (77.5%)	0.175 (73.1%)	495
IEnKS–SDA (lag 1)	10.272 (33.6%)	0.142 (40.7%)	800
IEnKS–MDA (lag 3)	10.078 (31.1%)	0.121 (19.9%)	9802
En4D–Var	10.210 (32.7%)	0.118 (16.6%)	611

Table 2.2: Comparing the accuracy of estimation and run time with 3 different initial guesses for the material parameters G and μ. Runs were performed on a computer with dual 12-core 2.3 GHz processors. Table reprinted from Spratt, Rodriguez, Schmidmayer, et al. (2021) with permission from Elsevier, © 2021.

Table 2.2 shows that with a relatively good initial guess with 10% error, the assimilation methods perform adequately. For example, the EnKF tracks the correct values for shear modulus and viscosity within 6% and 3%, respectively. With a moderate initial error of 50%, however, the EnKF loses accuracy and barely improves on the initial guess. In some cases, the EnKF was observed to be unstable, and the initial ensemble covariance had to be limited to prevent divergence. This limited the ability of the filter to estimate the parameters of interest, and thus despite its computational efficiency, the EnKF is eliminated from further consideration for this application.

The IEnKS and En4D–Var performed better than the EnKF for the 50% error case. The estimation was stable while varying initial conditions and covariance. Still, the lag 1 IEnKS–SDA only resulted in marginal improvements in the parameter values. The lag 3 IEnKS–MDA, on the other hand, resulted in further improvement, but at a high computational cost. This cost is associated with the calculation of the Hessian (see equation (2.29)) and gradient of the cost function (see equation (2.30)), which now involves three future time steps. The En4D–Var performs best in this test case, achieving good estimation with a comparably fast computational time. We note that while the En4D–Var was run for fifteen iterations in each case, the material property estimation converged by the fifth iteration. Thus, results and run-time after five iterations are reported.

Estimation results in the case with 50% error are presented in figure 2.3. Figure 2.3d shows the suitability of the En4D–Var: both parameters converge to accurate estimates within a few iterations. Overall, figure 2.3 also highlights the value of looking over a time horizon. While the EnKF and lag 1 IEnKS appear to disbelieve the data too much throughout the run, taking into account multiple times enables the lag 3 IEnKS–MDA to adjust to new information well, notably around collapse. Additionally, by nature, the IEnKS–MDA assimilates over the same data points multiple times. Data at any given time-step k, including near collapse point, will be used to minimize the cost functions at three separate times $k - 4$ to $k - 1$ for a lag 3 IEnKS–MDA. This redundancy appears important to parameter estimation in this context and given this data capture rate. Indeed, the IEnKS–MDA significantly corrects the viscosity estimate around each collapse, and the shear modulus estimate during the second collapse. Assimilating data from single time-steps appears to be insufficient given the short time scales of bubble cavitation and limited data. Smoothing over multiple times far improves performance around collapse points, which, given the IEnKS–MDA results, appear to hold the most pertinent data to make the necessary corrections.

In the case with 100% error in the initial guess, the relative performances of each method are similar to the 50% error case, but the three smoothers stagnate at 20 to 40% errors for μ and G. Weighted by computational expense, the En4D–Var performs best, but the IEnKS–MDA should not be discarded. Indeed, the time-varying estimation provides additional information about potentially time-dependent modeling uncertainties. While the physical model used assumes a constant shear modulus

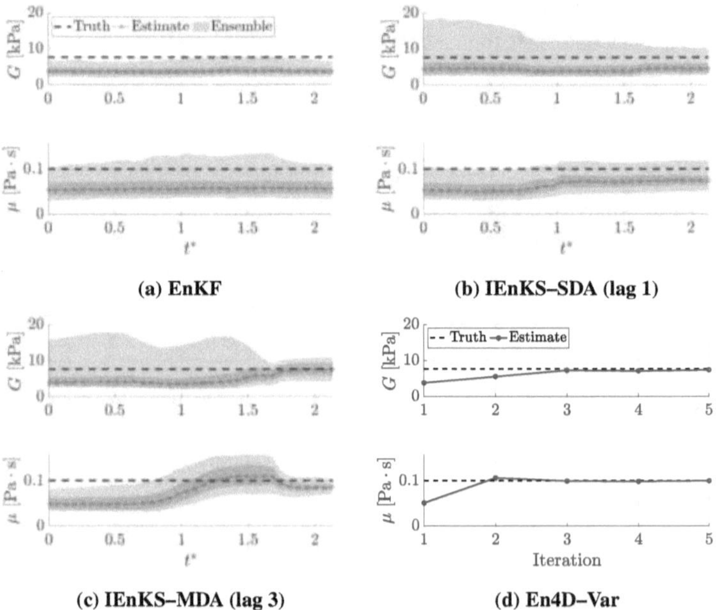

Figure 2.3: Estimation of shear modulus G and viscosity μ with initial guesses of $G = 3.8\,\text{kPa}$ and $\mu = 0.05\,\text{Pa s}$ (both at 50% error) in cases with surrogate experimental data (simulated radius time histories with added noise). The estimation is plotted over non-dimensional time t^* for the EnKF and IEnKS methods, and over iteration number for the En4D–Var. The dashed black line is the true parameter value, the blue circles are the estimate, and the blue shaded area is the ensemble spread. Reprinted from Spratt, Rodriguez, Schmidmayer, et al. (2021) with permission from Elsevier, © 2021.

and viscosity, the quasi-online IEnKS–MDA can uncover potential limitations of this assumption. This issue is further examined in section 3.1.

2.3.2 Uncertainty

Ensemble methods carry information about error statistics of the estimated parameters in the final ensemble. One way to visualize ensembles is through a histogram, an example of which is shown in figure 2.4 for the logarithm of the Cauchy number with the lag 1 IEnKS–SDA estimator. Despite the nonlinearity of the model, the tested methods track only the first two statistical moments of an assumed Gaussian filtering or smoothing PDF. Previous works (e.g., Evensen and Leeuwen (2000), Yang, Kalnay, and Hunt (2012), and Katzfuss, Stroud, and Wikle (2016)) have discussed that adequate results can still be achieved with a nonlinear model where this assumption must break down to some degree. Our results for the IEnKS and En4D–Var results above confirm that this is the case in this example.

Figure 2.5 shows a comparison of the fitted histograms for the methods for the case with 50% initial error in both parameters. Despite imperfect estimation, the En4D–Var converges significantly better than other methods given the limited data. The IEnKS–MDA curve displays the least variance of the Kalman methods, as expected.

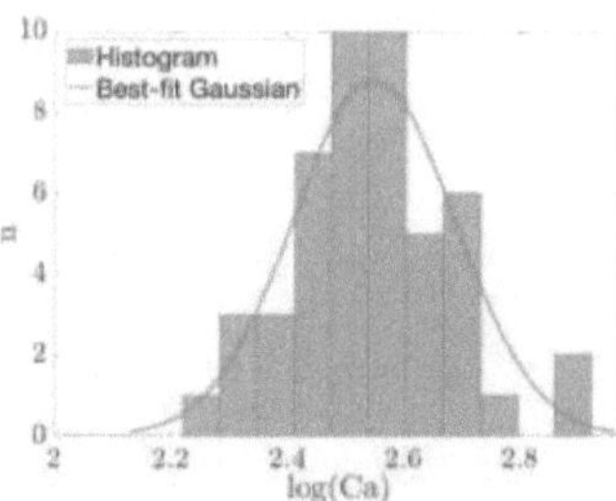

Figure 2.4: Histogram of the final estimate for log(Ca) with the lag 1 IEnKS and fitted normal curve, where n is the number of ensemble members at each value of log(Ca). Reprinted from Spratt, Rodriguez, Schmidmayer, et al. (2021) with permission from Elsevier, © 2021.

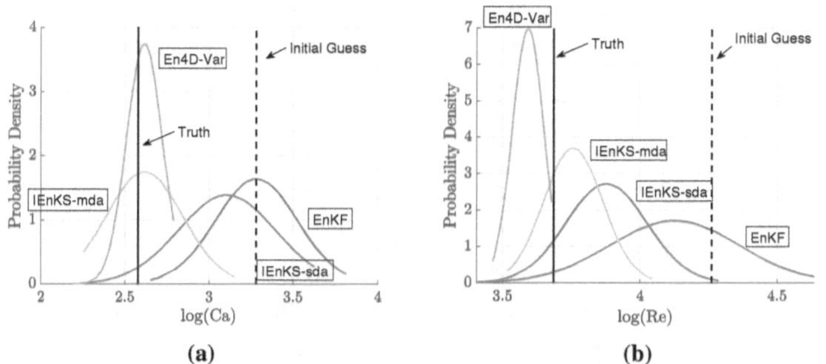

Figure 2.5: Comparison of final ensembles for log(Ca) (a) and log(Re) (b) estimates with each tested DA method in the case with 50% initial error in both parameters. Reprinted from Spratt, Rodriguez, Schmidmayer, et al. (2021) with permission from Elsevier, © 2021.

Anticipating experimental results where 10 data sets are available, the En4D–Var was run 10 times with the same ground truth but different (random) noise. Results are shown in figure 2.6 for shear modulus and viscosity estimates over the data sets. The dashed black lines correspond to the truth, and the blue line to the mean estimate over the 10 runs. Results across these 10 data sets are fairly uniform (standard deviation of 0.98 kPa for G, 0.009 Pa s for μ), confirming that reliable estimates are obtained despite noisy measurements across data sets. Figure 2.7 shows a histogram combining final ensembles for shear modulus to visualize overall results. As each of the 10 ensembles should be approximately normal, a Gaussian curve is expected when combining them. Figure 2.7 indeed shows an approximately normal distribution, as does the equivalent histogram for viscosity (as shown in section 3.1.3 in figure 3.3a).

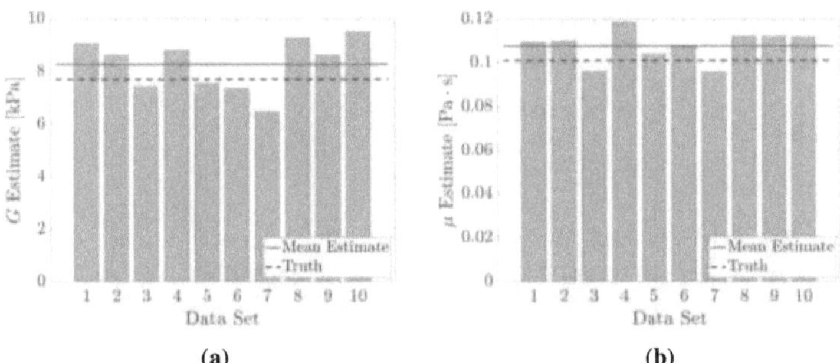

Figure 2.6: En4D–Var estimation results for G (a) and μ (b), for ten simulated data sets. Reprinted from Spratt, Rodriguez, Schmidmayer, et al. (2021) with permission from Elsevier, © 2021.

Based on these results with simulated data, given reasonable initial guesses as to the shear modulus and viscosity, we can confidently expect to estimate both parameters to within 5% using the 10 available data sets. Multiple initial guesses can be tested and their fits with experimental radius histories compared. In practice, an iterative process can be used to formulate a good initial parameter guess. That is, the final estimates of a first data assimilation run can be used as an initial guess for the next, and so on until results converge or an adequate radius fit is achieved. As shown in table 2.2, even with a large error, both IEnKS methods and the En4D-Var significantly improve on the initial guess, thus only a few iterations suffice to obtain a good estimate. The En4D-Var is particularly well suited for this, as it

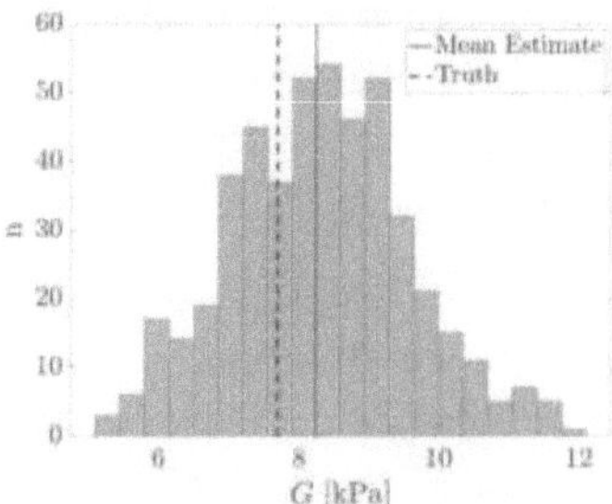

Figure 2.7: Histogram for G estimates combining 10 final ensembles for simulated data runs with En4D-Var. Reprinted from Spratt, Rodriguez, Schmidmayer, et al. (2021) with permission from Elsevier, © 2021.

remains stable even with large initial error and a wide spread in the ensemble, as long as all parameters retain physical values across the ensemble. Therefore, it is straightforward to formulate an initial guess with less than 50% error, and thus obtain results comparable to the second test case with simulated data. When applying this to experimental data, the En4D–Var serves as the baseline given its performance. IEnKS–MDA can also be tested for its ability to estimate parameters quasi-online.

We note that in all presented results, values for the β_i coefficients in the IEnKS–MDA and En4D–Var methods, which attribute weights to different time steps (see equation (2.23)), were held constant across all time-steps. Different values were tested for these coefficients to observe their impact on estimation results. For the IEnKS–MDA, we attempted to increase or decrease the coefficients as we stepped forward over the data assimilation window (linearly with varying slope). Neither approach, across all test cases, had a positive impact on the estimation. Thus we kept constant values for the coefficients, with $\beta_i = \frac{1}{3}$ for each $i \in [0, 3]$ for the data assimilation window of 3 time steps used.

For the En4D–Var methods, given the large amount of information gained near the collapse points during estimation, we attempted to increase the β_i coefficients near collapse points. That is, we posited that by increasing the weight attributed to data near collapse, and increasing the sensitivity of our estimates to collapse points, we may improve the accuracy of results. However, as with the IEnKS–MDA, no improvement in parameter estimation was observed given this strategy. Furthermore, for consistency, modifying coefficients in this way would require additional parameter tuning for each data set, as collapse occurs at different times in each. Hence, as with the IEnKS–MDA method, we pick constant coefficients of $\beta_i = \frac{1}{N}$ for the En4D–var, with N the total number of time-steps. This yields good results, without any added tuning required for β_i coefficients.

Based on the results presented in this chapter, DA-IMR—in particular the IEnKS–MDA and En4D–Var methods—appears to be a powerful framework for material parameter estimation, expanding on the capabilities of IMR and maintaining high accuracy. However, there are limitations to working with synthetic data. While noise with a high variance is added to the surrogate data, this is white noise added to data generated with the same model used in our DA forecast model. This means that the model used throughout this section matches exactly with the data (except for the added noise), which is a significant simplification of real-world cases, where the model carries uncertainty associated with its assumptions (as listed in section 2.1). We have shown that if the model used is perfect, we can accurately estimate material parameters even with high levels of noise, but this does not account for systematic modeling errors that could be present in experimental data. In the next chapter, we address this by applying DA-IMR to experimental data to further test its performance. Given the results from this chapter, we expect DA-IMR to perform well, as long as the physical model we use accurately captures the physics of bubble collapse in the hydrogels used.

Chapter 3

APPLICATION OF DA-IMR TO LASER AND ULTRASOUND-INDUCED CAVITATION

Parts of this chapter are adapted from Section 5 of Spratt, Rodriguez, Schmidmayer, et al. (2021), Chapters 4-5 of Buyukozturk et al. (2022) and Chapters 2-4 of Mancia, Yang, et al. (2021). This chapter presents applications of our developed DA-IMR framework in three separate cases with varying cavitation methods, material properties, experimental setups, and using different material models. This establishes the accuracy and versatility of our method, and provides concrete applications of the framework. We begin by applying the data assimilation methods found to be optimal in Chapter 2 to experimental laser-induced cavitation data, where bubble radius time-histories are obtained through high-speed imaging of cavitating bubbles in a gel subject to a high amplitude laser pulse. Here, we show that our method accurately retrieves material parameters with experimental data which the surrogate data used in Chapter 2 was designed to mimic. We discuss these experimental results in detail and compare them to previous results obtained with the same data sets. Then, we present an application with laser-induced cavitation in gels with nucleation seed particles. These particles enable cavitation with lower laser energy, to access new finite deformation and material stretch regimes. A new, more robust material model is also used here requiring estimation of new model parameters. We show the advantage of parameter estimation at lower energy, and use the DA methods to pinpoint uncertainties in our physical model. Finally, we present results of our DA framework in ultrasound-induced cavitation data. Using ultrasound reduces modeling uncertainty seen with laser-induced cavitation, as no plasma is formed in the material during cavitation. This highlights a new setting where DA-IMR can be used.

3.1 DA-IMR with Laser-induced Cavitation Data

In Chapter 2, all presented results were obtained with simulated data. There, the model used in our DA framework exactly matches that of the data, as the same forecast function in our DA methods is used to generate this data. This, of course, represents an idealization of experimental scenarios, as the model used may contain intrinsic error. Thankfully, these DA methods are well suited to address both experimental and model error, and we will show that we can predict material properties with a high degree of accuracy. In the absence of known 'true' material parameters, we will use the discrepancy between simulated radius vs. time curves and experimental data as the main metric to determine the adequacy of our results.

3.1.1 Experimental Setup

We begin by briefly describing the experimental setup for the first test case examined in this chapter, the data from which is used throughout section 3.1. This data was presented in Estrada et al. (2018), where a more detailed description of the setup for data collection can be found. Here, a polyacrylamide is used as the medium for cavitation, the properties of which will be estimated. After the polyacrylamide gel is prepared, each cavitation event is induced with a 6ns pulse of a "user-adjustable 1–50 mJ, frequency-doubled Q-switched 532 nm Nd:YAG laser." These cavitation events are triggered at different locations in the same large batch of polyacrylamide to maximize uniformity of material properties across experiments. Bubble radius is captured approximately every 3.7 µs, processing 270 000 fps high-speed camera output by subtracting a reference image from each frame and fitting a circle. A few sources of error may be present. Nonuniformity of the polyacrylamide gel or discrepancies across data sets could cause the bubble to lose spherical symmetry. However, each was triggered at least 5 maximum bubble radii away from a previous location or edge of the gel, to prevent any boundary effects which could introduce non-sphericity, for example through the formation of microjets. Estrada et al. (2018) report that the maximum $\dot{R}/c$ was approximately 0.4, and thus below a regime where significant non-spherical effects may be introduced during the initial collapse (Brujan, Nahen, et al., 2001; Sagar and el Moctar, 2020). Laser pulses may also vary slightly across runs, affecting the energy deposited in the system and thus initial growth conditions. In practice, a difference in maximum bubble radii was observed, with $R_{\max} = 388 \pm 35$ µm across experiments with the stiff gel, and $R_{\max} = 430 \pm 17$ µm with the soft gel. Ten experimental data sets in the stiff gel were used for the following results, in part to address this potential lack of uniformity in experimental conditions. As with the test case with surrogate data in Chapter 2, we obtain similar results for the stiff and soft gels, and report only results with the stiff gel here

3.1.2   En4D–Var Estimates for Shear Modulus and Viscosity

The noise magnitude in the experimental data is smaller than what was used in the simulated data with the same data rate. Therefore, if the model is adequate and the noise accurately represented as Gaussian, the IEnKS and En4D–Var should yield comparable or better estimation results with the experimental data. As with the simulated data, the assimilation window is limited to the initial collapse and two subsequent rebounds to match the setup used by Estrada et al. (2018), who estimated in the stiff-gel: $G_{stiff} = 7.69 \pm 1.12$ kPa and $\mu_{stiff} = 0.101 \pm 0.023$ Pa s. The results are compared to theirs. Our estimation is initialized with three different initial guesses, detailed in table 3.1. Similarly to the surrogate truth data from section 2.3.1, initial guesses with 10%, 50%, and 100% difference from the Estrada et al. (2018) estimates are chosen.

Method	G estimate $\pm\sigma$ [kPa]	μ estimate $\pm\sigma$ [Pa s]	Run time [s]
Guess 1	8.50 (+10% diff)	0.09 (-10% diff)	–
IEnKS–SDA (lag 1)	7.93 ± 1.68	0.096 ± 0.012	2751
IEnKS–MDA (lag 3)	7.51 ± 1.50	0.089 ± 0.016	9536
En4D–Var	7.41 ± 1.63	0.093 ± 0.014	609
Guess 2	3.80 (-50% diff)	0.05 (-50% diff)	–
IEnKS–SDA (lag 1)	4.32 ± 0.46	0.085 ± 0.013	2832
IEnKS–MDA (lag 3)	6.67 ± 1.43	0.083 ± 0.016	10052
En4D–Var	6.53 ± 1.58	0.090 ± 0.014	585
Guess 3	15.0 (+100% diff)	0.20 (+100% diff)	–
IEnKS–SDA (lag 1)	9.46 ± 2.76	0.114 ± 0.014	2871
IEnKS–MDA (lag 3)	8.57 ± 1.52	0.103 ± 0.015	9222
En4D–Var	8.24 ± 1.58	0.098 ± 0.016	535

Table 3.1: Comparing results of estimation with three different initial guesses for the material parameters G and μ. Runs were again performed on a computer with dual 12-core 2.3Ghz processors. Table reprinted from Spratt, Rodriguez, Schmidmayer, et al. (2021) with permission from Elsevier, © 2021.

Table 3.1 summarizes the results from the three different initial material parameter guesses for the three methods. These estimates correspond to the mean estimate over all 10 experimental data sets. The standard deviation σ of the results is also reported.

The En4D–Var estimates for shear modulus and viscosity all fall within the error bounds provided by Estrada et al. (2018). Results are close to theirs in the test cases considered. The estimates are uniform, with only 8.5% difference between viscosity estimates, and 23% difference in the shear modulus results across the data sets. This larger difference in the shear modulus estimates and in the associated standard deviations is expected, as we have found the radius curves to be relatively more sensitive to μ than G. Finally, the average normalized root mean squared error (NRMSE) for bubble radius is low at 2.16×10^{-2} for guess 1, indicating that a good fit was achieved with this method. This NRMSE is obtained by re-running the simulation with the forecast model and final parameter estimates, and comparing the results to experimental data using equation (3.1). Given that the guess–1 results lead to the smallest radius error, our initial shear and viscosity modulus estimates are $G = 7.41 \pm 1.63\,\mathrm{kPa}$ and $\mu = 0.093 \pm 0.014\,\mathrm{Pa\,s}$, respectively. An example bubble radius curve is shown in figure 3.1, for one of the experimental data sets (data set 10).

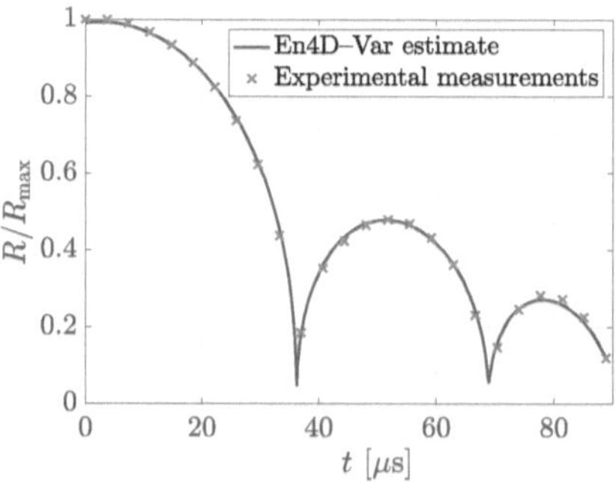

Figure 3.1: Simulated radius vs. time curve given by En4D–Var estimates for G and μ, and experimental measurements for data set 10. Reprinted from Spratt, Rodriguez, Schmidmayer, et al. (2021) with permission from Elsevier, © 2021.

The standard deviation across the 10 IEnKS–SDA runs was comparable to the En4D–Var. However, the estimates varied significantly based on the initial guess: from $4.32\,\mathrm{kPa}$ to $9.46\,\mathrm{kPa}$ for shear modulus, and from $0.085\,\mathrm{Pa\,s}$ to $0.114\,\mathrm{Pa\,s}$ for viscosity. Except for estimates from guess 1, the shear modulus estimates are also outside the bounds given by Estrada et al. (2018), and the radius fit is significantly worse than that of the En4D–Var, with an average NRMSE of 9.10×10^{-2}.

On the other hand, while still worse than that of the En4D–Var, the IEnKS–MDA estimates are within the Estrada et al. (2018) margin, and the radius fit is better than that of the IEnKS–SDA (NRMSE $= 6.79 \times 10^{-2}$). The IEnKS–MDA thus represents the best tested quasi-online method, as expected from the simulated data results of section 2.3.1. It is important to note here that the bubble radius fits were all obtained by re-running simulations with final shear modulus and viscosity estimates and comparing to experimental measurements. This is a fair way to compare the ability of each method to estimate these parameters. However, the radius fit obtained online during assimilation with the IEnKS methods is better (and comparable with the En4D–Var estimates), given that the radius is also being directly corrected at each time-step as part of the state vector. For parameter estimation, the En4D–Var is the best tested method, but the IEnKS–MDA is a good quasi-online estimator. This is particularly useful for the discussion in section 3.1.3, where we make use of this time-varying estimation.

3.1.3 Accounting for Model Uncertainty

The estimates obtained for the shear modulus and viscosity from the previous section show that ensemble data assimilation methods can be effectively used for estimation of viscoelastic material properties. A further look at the results, though, provides more information than simply this estimate. Examining estimates for each variable across the 10 tested experimental data sets, as shown in figure 3.2, there appears to be a discrepancy between data sets 3, 4, 5 and the rest for the viscosity. While the shear modulus estimation shows no discernible trend (despite the previously mentioned larger spread in results), the viscosity data appears to be split between two estimates. The red line in figure 3.2b shows the mean estimate of data sets 3 to 5 ($\mu = 0.074\,\mathrm{Pa\,s}$) and the green line that of the rest ($\mu = 0.102\,\mathrm{Pa\,s}$).

Figure 3.3 compares the histogram obtained when collating the $10 \times q$ final ensemble members for viscosity from 10 runs with different simulated data but the same ground truth (Figure 3.3a) and the 10 runs done with experimental data (Figure 3.3b). As discussed in section 2.3.2, we expect to approximately retrieve a Gaussian distribution around the estimate, as is the case for the simulated run in figure 3.3a. However, figure 3.3b shows an apparent bimodal distribution. The lower viscosity peak corresponds to the mean estimate of data sets 3 to 5, and the higher peak to that of the rest.

To understand what may be causing this discrepancy in results, it is useful to consider the IEnKS–MDA and its quasi-online estimation of viscosity. Figure 3.4 shows a

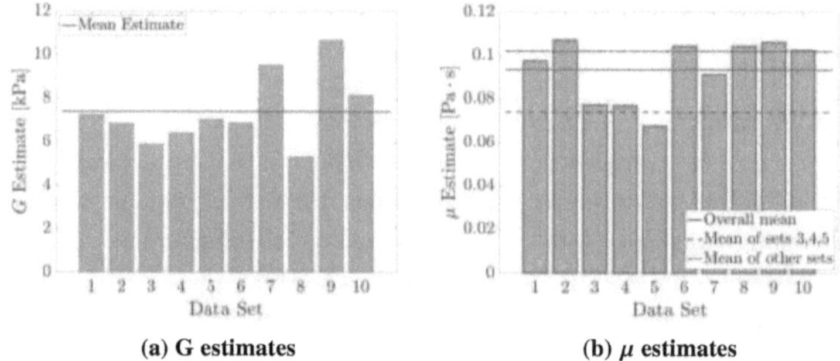

(a) G estimates   (b) μ estimates

Figure 3.2: En4D–Var estimates for shear modulus (a) and viscosity (b) for 10 experimental data sets. Reprinted from Spratt, Rodriguez, Schmidmayer, et al. (2021) with permission from Elsevier, © 2021.

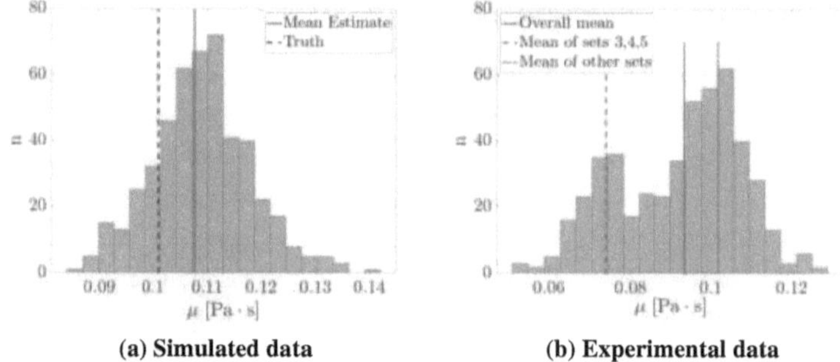

(a) Simulated data   (b) Experimental data

Figure 3.3: Comparison of final combined ensembles for viscosity estimation in simulated and experimental data. Reprinted from Spratt, Rodriguez, Schmidmayer, et al. (2021) with permission from Elsevier, © 2021.

comparison between viscosity estimation for data set 2 (figure 3.4a) and data set 3 (figure 3.4b). These are representative of data sets with a high and low viscosity estimate, respectively. They result in estimates of $\mu = 0.098$ Pa s for data set 2, and $\mu = 0.077$ Pa s for data set 3. Comparing the two data sets, it appears that the assimilation begins similarly, correcting to a higher viscosity estimate during the first collapse. However, there is a divergence between the behavior of the smoother after each collapse point, particularly the second one (around $t = 65\,\mu$s). Figure 3.4a shows a slow decrease and convergence towards a higher viscosity value, with negligible change at the second collapse point. However, this estimate drops sharply after these collapse points in figure 3.4b. In fact, the data around each collapse causes the viscosity estimate to sharply drop in data set 3, which does not occur in data set 2. This behavior is representative of what is seen in data sets 3 through 5, but does not occur in the rest of the runs.

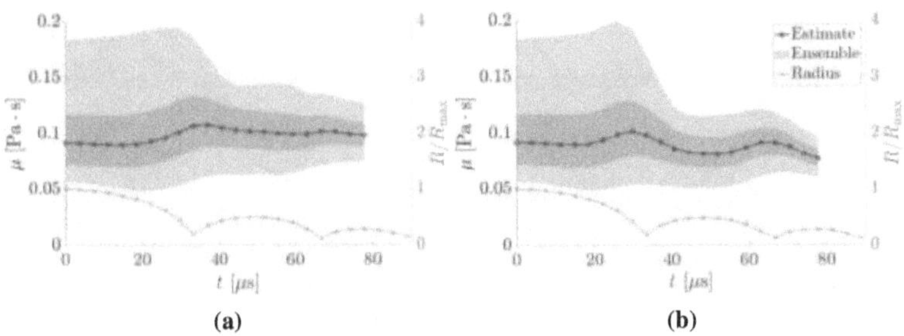

Figure 3.4: Comparison of online estimation of viscosity in data sets 2 (a) and 3 (b) using the IEnKS–MDA. Reprinted from Spratt, Rodriguez, Schmidmayer, et al. (2021) with permission from Elsevier, © 2021.

Given the physical model used, the viscosity should be constant and such drops in the parameter are not expected. The model alone thus cannot adequately capture the behavior of the gel seen by the IEnKS in these data sets. We can posit that a violent collapse in these data sets is causing inelastic behavior in the material, and thereupon this perceived change in material properties (Yang, Cramer, and Franck, 2020). More work will be needed to determine the exact cause, but this could perhaps result from fracture, damage to the polymer network in the gel, or combustion in the gas phase (Movahed et al., 2016; Kundu and Crosby, 2009; Raayai-Ardakani, Earl, and Cohen, 2019).

To further examine this, we perform two simulations where we slightly modify our spherical bubble dynamics model. Here assumption of constant temperature in the surrounding material is relaxed, and we apply conservation of energy to obtain a partial differential equation for this temperature field. Appendix A describes the changes to the model used to calculate material temperature, based on the model of Preston, Colonius, and Brennen (2007). Specifically equation (A.10) is used for temperature in the surrounding medium. The material temperature field T_m is added to the state vector x, and estimated along with the temperature field inside the bubble and all other variables. While it has predominantly been used to estimate constant model parameters G and μ so far, IEnKS provides quasi-online estimates of all state variables, which now include the temperature field in the surrounding gels. Figure 3.5 shows a comparison of heat-maps of this estimated temperature field for data sets 2 and 3, where the x-axis represents non-dimensional time $t^* = \frac{t}{R_{max}}\sqrt{\frac{p_\infty}{\rho}}$, the y-axis the distance from the bubble wall, and the color of each cell is non-dimensional temperature.

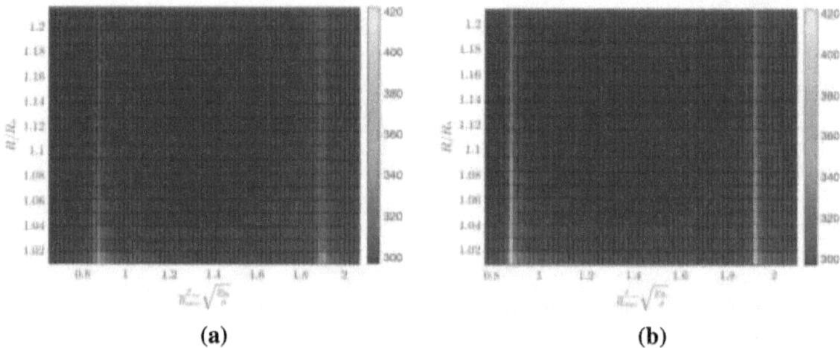

Figure 3.5: Comparison of online estimation of the temperature field in the surrounding material for data sets 2 (a) and 3 (b) using the IEnKS–MDA. The x-axis is non-dimensional time, the y-axis represents distance from the bubble wall, and the color of each cell represents temperature.

In both cases, temperature peaks are visible at two distinct times, which correspond to the first and second collapse of the bubble. However, the magnitude of these temperature peaks is larger for data set 3, the same data set which saw the sharp drop in viscosity estimate at collapse time. This could indicate that the violence of collapse in this data set results in high temperatures at the bubble wall, which in turn causes damage to the material. Regardless of physical cause, this time-dependent behavior is not accounted for in the model, but is captured by the IEnKS–MDA as a drop in the perceived viscosity. While the exact bounds of the physical model used are not known and would require more data to determine, this shows that in these particular data sets, model accuracy is far reduced after the first collapse.

Figure 3.6 compares the normalized root mean squared error (NRMSE) across all data sets, given by

$$\text{NRMSE} = \sqrt{\frac{1}{N}\sum_i^N \frac{\left(R_{\text{sim}}(t_i) - R_{\text{exp}}(t_i)\right)^2}{R_{\text{exp}}(t_i)^2}} \tag{3.1}$$

where R_{exp} is the experimental bubble radius time history, and R_{sim} is the simulated time history given the estimated material properties (at the corresponding times).

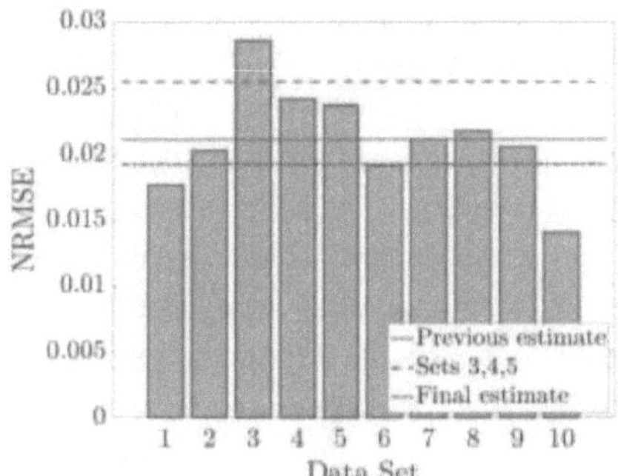

Figure 3.6: Bar plot of radius normalized root mean squared errors (NRMSE) for each data set. Also plotted are the previous estimate mean NRMSE (mean of all sets), the mean NRMSE for sets 3 to 5, and the final estimate mean NRMSE (mean of all other sets). Reprinted from Spratt, Rodriguez, Schmidmayer, et al. (2021) with permission from Elsevier, © 2021.

Figure 3.6 shows a higher error in the estimated bubble radius curves fit for data sets 3 through 5, which is expected given the heightened model uncertainty in these data sets. Because of this uncertainty and higher error, we discard these three sets as outliers, which yields the final IEnKS–MDA-informed En4D–Var estimate reported in table 3.2. Notable is the drop in standard deviation for viscosity as compared to the previous En4D–Var estimate and the reduced NRMSE.

Estimate	$G \pm \sigma$ [kPa]	$\mu \pm \sigma$ [Pa s]	NRMSE
Estrada et al. (2018)	7.69 ± 1.12	0.101 ± 0.023	
Previous	7.41 ± 1.63	0.093 ± 0.014	2.16×10^{-2}
Final	**7.81 ± 1.80**	**0.102 ± 0.006**	1.95×10^{-2}

Table 3.2: Final En4D–Var estimates (discarding three outlier data sets) and standard deviation, along with the average radius normalized root mean squared error. The previous best estimate corresponds to the mean of all 10 data sets, outliers included. Table reprinted from Spratt, Rodriguez, Schmidmayer, et al. (2021) with permission from Elsevier, © 2021.

Given the similarity to the values obtained by Estrada et al. (2018) and the low NRMSE of the resulting radius time history simulations, these results show that ensemble-based data assimilation can successfully estimate the mechanical properties of hydrogels at these high strain rates. The En4D–Var method efficiently provides an accurate estimate of material properties. In addition, the IEnKS–MDA method can elucidate time-dependent behavior, even when not fully accounted for by the forecast model used, by estimating parameters quasi-online. These methods account for both model and experimental error, dealing with noisy measurements or inaccuracies in the model well. It is hypothesized here that the bubble collapses are damaging the polyacrylamide gel in certain test cases, leading to a reduced estimated viscosity after each subsequent collapse, a physical effect not accounted for in the model. Finally, added benefits of these algorithms include adaptability to different numerical or viscoelastic models, and scalability to further parameter estimation, with negligible computational cost for additional parameters. The following chapters will demonstrate these benefits, as we apply these methods to new material models and estimate additional parameters.

3.2 DA-IMR with Laser-induced Cavitation Data in Seeded Gels

In the results of section 3.1, we applied our DA framework to experimental data sets within the same deformation regime. The material stretch ratio, defined (assuming a spherical bubble) as the maximum value of the hoop stretch over the bubble collapse

$$\lambda_{\max} = R_{\max}/R_0, \tag{3.2}$$

where R_0 is the long-time bubble equilibrium radius, varies little across experiments. This introduces some limitations to the scope of the DA-IMR method. Namely, across various materials and applications, maximum deformation can vary broadly. The breakdown process undergone during laser-induced cavitation is highly dependent on the material's properties, and can lead to drastically different stretch ratios in different materials. Thus, testing our characterization framework in different stretch ratio regimes would inform how broadly it can be applied. To address this, Buyukozturk et al. (2022) introduce a novel experimental approach, whereby the stretch ratio can be controlled within the same material by introducing seed particles and varying laser energy. The seed particles are used as a nucleation site to cavitate bubbles, reducing the dependence of bubble dynamics on properties of the material itself. Indeed, the particles make it far easier for cavitation to occur, and lower laser energy can be used. Varying both laser energy and the nature of the seed particles, a wide range of stretch ratios can be achieved for a single material.

Furthermore, it is known that the bubble dynamics physical model used, described in section 2.1 in Chapter 2, can be limited in the case of extreme loading rates (Gent and Wang, 1991; Barney et al., 2020; Hashemnejad and Kundu, 2015; Hutchens,

Fakhouri, and Crosby, 2016). As discussed in section 3.1.3, the model does not account for material fracture or other failure mechanisms, and assumes a consistent viscoelastic response. By using our data assimilation approach, we can continue to quantify modeling uncertainty and identify regimes or critical points in the bubble collapse where our models may be failing to adequately capture the physics. Applying DA-IMR to the experimental method of Buyukozturk et al. (2022), which extends stretch ratio regimes where cavitation can occur, will much more broadly assess the adequacy of our DA-IMR method in determining material properties of viscoelastic materials.

3.2.1 Experimental Setup and Results

We briefly describe the experimental setup used and results obtained by Buyukozturk et al. (2022), to which we apply our DA-IMR framework in section 3.2.2. Overall, the setup is similar to that of Estrada et al. (2018), used to generate the data analyzed in section 3.1. The hydrogel used is again polyacrylamide, with a quasi-static shear modulus of $G_\infty = 461 \pm 4\,\text{Pa}$. Four samples are prepared. In three of them, seed particles are distributed throughout the gels, respectively microspheres of glass, stainless steel, and paramagnetic coated polyethylene. For details on the gel preparation and laser setup used to nucleate bubbles, see section 2 of Buyukozturk et al. (2022). The key differences with the setup used to generate the data of section 3.1 and shown in figure 1.1 are that the laser will be focused on seed particles (in gels containing them) to initiate cavitation, and the use of neutral density (ND) filters to attenuate the laser energy, thus modulating the energy between $18.4 \pm 1.0\,\mu\text{J}$ and $449.2 \pm 5.0\,\mu\text{J}$. Finally, the high-speed camera used to image bubbles is capable of far greater frame rates than in the previous setup. In the data sets used in the following, bubble radius is captured every $1\,\mu\text{s}$, at 1 million frames per second (compared to 270,000 previously). We saw in section 3.1.3 when using the En4D-Var method that the estimator was sensitive to data around the first bubble collapse. However, the time-resolution of those experiments meant that only a few data points were available near this time, given the violence of the dynamics. Increasing the time-resolution of our data by almost a factor of 4 means we will be able to resolve the collapse much better, and improve material parameter estimation.

As expected, the use of seed particles enables cavitation across various stretch ratio regimes. Both particle type and laser energy have an impact on the maximum bubble radius. For cases with no seed particles, the stretch ratio hovered around $\lambda_{\max} = 9$, with values averaging $\lambda_{\max} = 9.2$ for the highest laser energy (449 µJ), and reducing to $\lambda_{\max} = 8.8$ at the lowest laser energy where cavitation can still occur with no seed particles present, around 118 µJ. In contrast, the stretch ratio can be reduced all the way to a mean of $\lambda_{\max} = 4.5$ in cases with the paramagnetic beads and laser energy 18.4 µJ. While the reduced laser energy does affect stretch ratio, the nature of seed particles has a large effect as well. For example, with a laser energy of 118 µJ, the stretch ratio is $\lambda_{\max} = 8.8$ in the gel without particles, $\lambda_{\max} = 5$ for paramagnetic particles, $\lambda_{\max} = 5.6$ for glass particles, and $\lambda_{\max} = 7.2$ for steel particles. Figure 3.7 summarizes the stretch ratios across all experimental data sets. We note that with seed particles present, laser energy was capped at 254 µJ to avoid cases where loss of sphericity was observed.

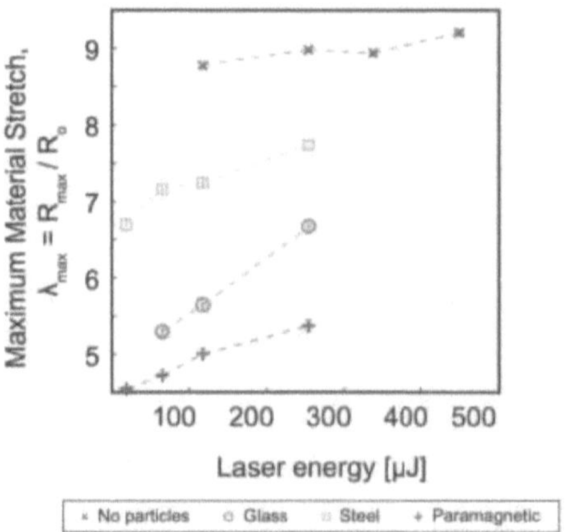

Figure 3.7: Value of Maximum stretch ratio $\lambda_{\max}$ vs. laser energy obtained across experiments with varying seed particles. All points represent the mean of 8 experimental data sets. Reproduced from Buyukozturk et al. (2022) with permission from Springer Nature, © 2022.

3.2.2 DA-IMR Parameter Estimation Results

We implement ensemble-based data assimilation techniques described in Chapter 2 and used in section 3.1 to fit the hydrogel's material properties and account for uncertainties in the physics model. In particular, the IEnKS-MDA and En4D-Var are used, with some minor modifications described here. First, some elements of the bubble dynamics model described in section 2.1 are improved. The Keller-Miksis equation (equation (2.1) (Keller and Miksis, 1980)) is still used, with the physics of a two-phase mixture of condensable water vapor and non-condensable gas in the bubble tracked, as shown to be appropriate in the literature (Estrada et al., 2018; Yang, Cramer, and Franck, 2020; Akhatov et al., 2001; Nigmatulin, Khabeev, and Nagiev, 1981; Barajas and Johnsen, 2017; Vincent et al., 2014). The same assumptions described in section 2.1 are maintained, however, the viscoelastic material model used is changed. Indeed, Yang, Cramer, and Franck (2020) showed that the quadratic law Kelvin-Voigt model, an adaptation of the Fung model (Fung, 2013) where higher order strain stiffening terms are used (Movahed et al., 2016), is best for modeling very high strain rate behavior of gels such as the presently used polyacrylamide. The use of this more general model may help estimation as we expand DA-IMR to new material stretch regimes, as compared to results from section 3.1.

Rather than estimating a shear modulus G as done in section 3.1, we fix a quasi-static shear modulus G_∞ and estimate a strain-stiffening parameter α. The stress integral used with this model is then defined as

$$
\begin{aligned}
S_{Quad} = {} & \frac{(3\alpha - 1)G_\infty}{2}\left[5 - \left(\frac{R_o}{R}\right)^4 - \frac{4R_o}{R}\right] - \frac{4\mu\dot{R}}{R} \\
& + 2\alpha G_\infty\left[\frac{27}{40} + \frac{1}{8}\left(\frac{R_o}{R}\right)^8 + \frac{1}{5}\left(\frac{R_o}{R}\right)^5 + \left(\frac{R_o}{R}\right)^2 - \frac{2R}{R_o}\right].
\end{aligned}
\tag{3.3}
$$

To account for these changes, the state vector used (previously equation (2.10)) becomes

$$
x = \{R, \dot{R}, p_b, S, \mathbf{T}, \mathbf{C}, \alpha, \mu\}.
\tag{3.4}
$$

Otherwise, the IEnKS and En4D-Var implementation of sections 2.2.2 and 2.2.4 remain unchanged. A lag of 3 is used with the IEnKS.

Now, in section 3.2.2.1, we perform bubble radius fits using the offline (En4D-Var) method to determine the Quadratic law Kelvin-Voigt material parameters up through the first (1-peak), second (2-peak), and third (3-peak) collapse. Material parameters are then used to calculate the critical Mach numbers for each initial bubble collapse using the IMR framework. For the online method, we assess the time-varying fitted material properties of the quasi-online (IEnKS) method to identify deviations from the model during the bubble time evolution in section 3.2.2.2. We note that the analysis of results of the following sections 3.2.2.1 and 3.2.2.2 was done in collaboration with Dr. Selda Buyukozturk and the other co-authors of Buyukozturk et al. (2022).

3.2.2.1 Multi-peak Fitting with the Offline (En4D-Var) Method

Numerical results presented in this section use the offline (En4D-Var) method for determining the viscoelastic material parameters in PA. Initial guesses into the solver for viscosity, μ, and strain stiffening parameter, α, are 0.05 Pa·s and 0.5, respectively. These are derived from a preliminary run with the former version of IMR using a least squares fitting scheme for fitting material properties Yang, Cramer, and Franck, 2020.

As a measure of the $R(t)$ goodness of fit given the En4D-Var parameter estimates, as with section 3.1, we use the normalized radius root mean squared error (NRMSE). Representative $R(t)$ fits with median NRMSE values for each fitted peak case are shown in Fig. 3.8. For a cavitation bubble in PA (paramagnetic seed particle), induced at nominal laser energy 117 μJ, the 1-peak case has a median NRMSE value of 0.009 (Fig. 3.8a), 2-peak case has a median NRMSE value of 0.027 (Fig. 3.8b), and the 3-peak a median NRMSE value of 0.047 (Fig. 3.8c).

We compare $R(t)$ curves obtained with the En4D-Var parameter estimates to experimental data to calculate the NRMSE. Figs. 3.8(d-f) compare λ_{max} to the NRMSE for each peak fit. As the number of fitted peaks increases, the NRMSE also increases. 1-peak fits (Fig. 3.8d) exhibit the lowest NRMSE, with most errors within 0.02. However, the error for the 2-peak fits (Fig. 3.8e) are mostly within 0.05, with some large scatter exhibited by the steel LIC cases. The 3-peak fit case (Fig. 3.8f) NRMSE values are mostly within 0.15, with a few outliers. Given the number of fitted peaks, stretch ratio does not have a noticeable effect on the NRMSE. However, with increasing number of peaks fitted, the NRMSE from the 1- to 3-peak fits increases by almost an order of magnitude. Thus, subsequent oscillatory bubble dynamics beyond the first peak are not as accurately described by the current theoretical IMR framework.

Next, we examine the effect of laser energy on NRMSE. In the 1-peak fit case (Fig. 3.8g), LIC bubbles in PA with no particles have the lowest NRMSE compared to the cases with PA with seed particles. While the general trend indicates increasing

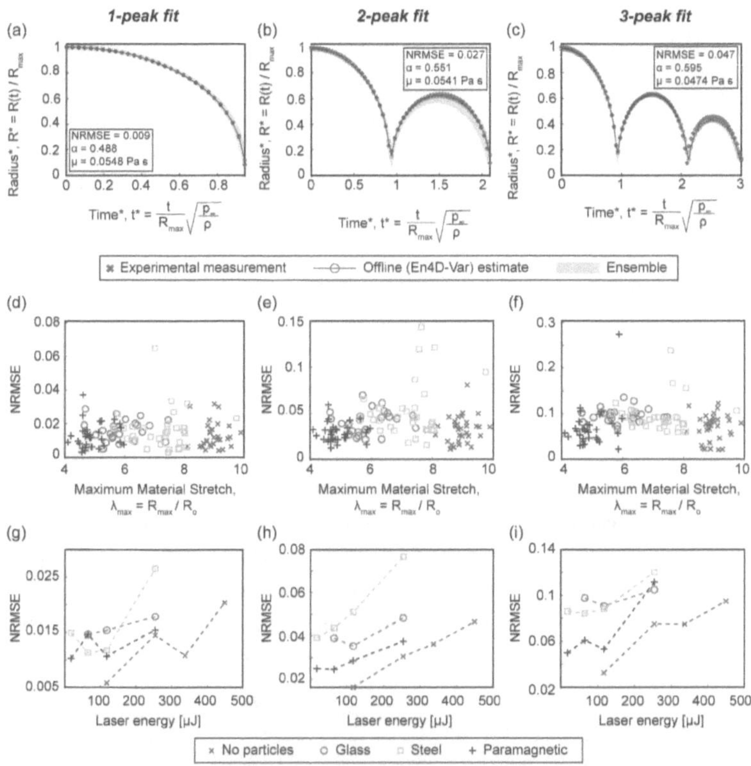

Figure 3.8: **Results of offline En4D-Var fitting on experimental curves for 1-, 2- and 3-peak fits. Representative examples of multi-peak fits with median NRMSE values (for PA paramagnetic particles at nominal laser energy 117 μJ) are shown for a (a) 1-peak fit (NRMSE = 0.009), (b) 2-peak fit (NRMSE = 0.027), and (c) 3-peak fit (NRMSE = 0.047). NRMSE values with respect to maximum material stretch for (d) 1-peak, (e) 2-peak, and (f) 3-peak fits, and with respect to laser energy for (g) 1-peak, (h) 2-peak, and (i) 3-peak fitting cases. Reproduced from Buyukozturk et al. (2022) with permission from Springer Nature, © 2022.**

NRMSE with increasing laser energy, the bubbles cavitated in PA with seed particles tend to have similar errors at the lowest energies and diverge more with increasing energy. This trend continues in the 2- and 3-peak fit cases, as shown in Figs. 3.8(h, i). Overall, increasing the laser energy increases the fit error.

For the range of samples and laser energies presented, the strain stiffening parameter, α, and viscosity, μ, were estimated using the En4D-Var with the IMR framework and Quadratic law Kelvin-Voigt stress integral (equation (3.3)). A summary of fitted material parameters for all cases is shown in Fig. 3.9, where the weighted-mean viscosity, μ, and strain stiffening parameter, α, values are plotted against each other with standard deviation error bars for each experimental condition. (The type of sample is denoted by symbol and the energy level is denoted by color, in accordance with the color bar shown.) The 1-peak fit case in Fig. 3.9a contains a large cluster of material parameters located within the range of $\mu \approx 0.04 - 0.08$ and $\alpha \approx 0.4 - 0.6$ with standard deviation values of $\sigma_\mu = 0.008$ and $\sigma_\alpha = 0.08$. A few higher energy cases fall outside of the cluster. However, as the number of fitted peaks increases to 2-peak (Fig. 3.9b) and 3-peak (Fig. 3.9c) fits, an increase in both μ and α values is observed, though the viscosity term is more affected. Another noticeable trend is that the standard deviation of fitted material parameters for a given experimental condition increases with respect to the laser energy. Thus, as energy increases, so does the spread in fitted material parameters, particularly for the 2- and 3-peak fitted cases. Notably though, parameter estimates for lower energy cases remain more clustered, and do not increase as much as more peaks are fitted. This may indicate that the IMR framework remains suitable over multiple peaks for lower laser energies.

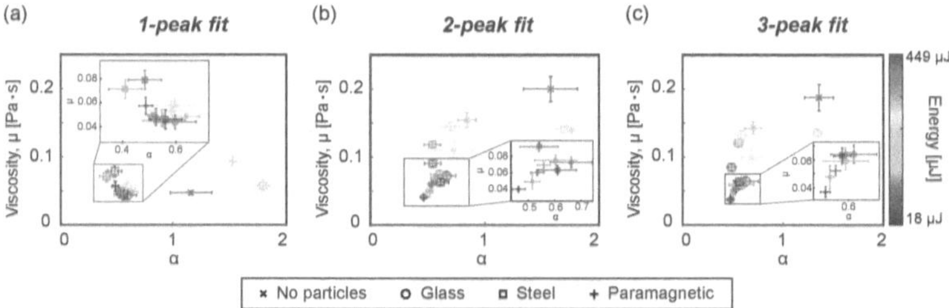

Figure 3.9: Offline En4D-Var estimates for Quadratic law Kelvin-Voigt material model parameters, namely viscosity μ, and strain stiffening parameter α. The material parameters estimates are shown for (a) 1-peak, (b) 2-peak, and (c) 3-peak fits. The symbols correspond to sample type and the color corresponds to laser energy, with the associated color bar shown on the far right. Each point is a weighted average of $n = 8$ experiments where error bars are standard deviation. Reproduced from Buyukozturk et al. (2022) with permission from Springer Nature, © 2022.

Upon collapse, inertial cavitation bubbles emit acoustic waves that propagate through the hydrogel matrix. The Mach number, M, over time is a function of the velocity of the bubble wall and the finite wave speed of the material, given by

$$M = \dot{R}/c. \tag{3.5}$$

The critical Mach number, M_{cr}, is the maximum Mach number of an inertial bubble, typically occurring at the first bubble collapse. Previous work suggests that M_{cr} values exceeding 0.08 define the condition for a violent collapse (Yang, Cramer, and Franck, 2020), which may introduce additional physical phenomena such as damage and inelastic material behavior. For more details on the IMR theoretical framework, as well as the derivation of non-dimensionalized expressions required for computations, see Estrada et al. (2018).

Given the estimated viscoelastic material parameters, the time evolution of bubble radii is simulated with the physical model equations, and the critical Mach number for each experiment is calculated using equation (3.5). In Fig. 3.10a, the critical Mach number, M_{cr}, with respect to each corresponding maximum stretch ratio, λ_{max}, is plotted for the 1-peak case and it is found that as the stretch ratio increases, so does the critical Mach number. The 2-peak (Fig. 3.10b) and 3-peak (Fig. 3.10c) cases follow the same trend, though there is an observable increase in scatter, particularly for the steel seed particle case. Thus, there seems to be a dependency of critical Mach number on the stretch ratio. Comparing the critical Mach number to laser energy, we refer to Fig. 3.7, where increasing energy only slightly increases the maximum stretch ratio. Given this, we can conclude that the critical Mach number is similarly minimally affected by laser energy.

The offline (En4D-Var) method applies the IMR framework to the experimental kinematic bubble radii using the Quadratic law Kelvin-Voigt material model to assess the goodness of the bubble radius history fit and viscoelastic material properties for up to 3-peak fits. The radius fit error, quantified as NRMSE, is generally lower for all cases at lower energy levels, regardless of the number of peaks fitted. However, the radius fit is worse, increasing by almost an order of magnitude, as we increase the DA window from 1- up through 3-peaks with respect to both maximum material stretch and laser energy. While the calculated critical Mach number did not have a noticeable change with respect to input laser energy, it was found that the critical Mach number increases with maximum material stretch ratio, a trend consistent regardless of the number of bubble radius peaks fitted. Lastly, the fitted viscoelastic

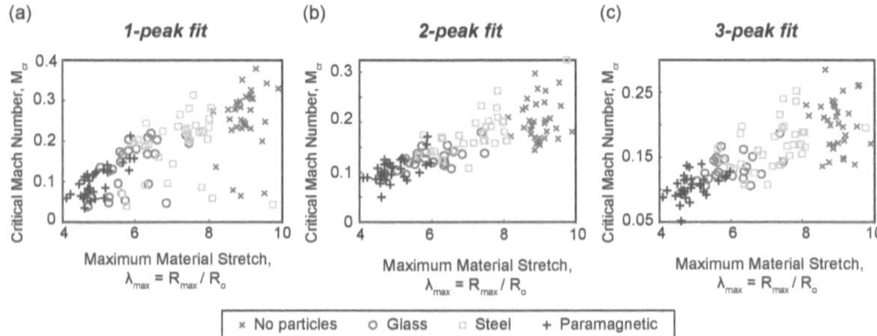

Figure 3.10: Critical Mach number, M_{cr} plotted with respect to maximum material stretch, λ_{max}, Using offline (En4D-Var) estimates given the (a) 1-peak, (b) 2-peak, and (c) 3-peak fitted solutions. Critical Mach numbers were obtained by re-running simulations given these En4D-Var estimates in each case. Reproduced from Buyukozturk et al. (2022) with permission from Springer Nature, © 2022.

material properties tend to cluster with a 1-peak fit, but the spread in mean values, as well as the standard deviation of values at a given experimental condition, increases with respect to higher laser energy and number of peaks fitted. The latter particularly affects the fitted viscosity parameter. Critical Mach number increases with respect to material stretch as before, leading to violent collapses at higher material stretches ($M \geq 0.08$).

3.2.2.2 Quasi-online (IEnKS) Results

Next, we implement the quasi-online (IEnKS) method, where the estimation trails the simulation by a fixed number of time steps. This is particularly useful because the time-varying estimation provides additional information on IMR model uncertainties. In this section, we will focus on two examples that indicate deviations from the theoretical framework, while comparing cavitation events nucleated under different experimental conditions.

For the first example, we focus on bubbles nucleated in PA (no seed particles) at all four laser energies, ranging from nominal values of 117 μJ to 449 μJ. In Fig. 3.11, we plot the ensemble's variance in these estimates over time. The time evolution of curves spans from the first bubble radius peak for all cases, where initial bubble collapse for each example occurs approximately at the normalized time indicated by the dashed vertical lines. For the nominal laser energy of 117 μJ case shown in Fig. 3.11a, the time-varying α parameter gradually increases then dips after the first bubble collapse. In Figs. 3.11(b,c), as the laser energy is increased, the α and μ variance have an increased peak. This is even more apparent at the highest laser energy case shown in Fig. 3.11d, where at a nominal laser energy of 449 μJ, both peak α and μ variances increase almost five times the peak values of the lowest energy case. Thus, as laser energy increases, the quasi-online (IEnKS) method produces more uncertain fitting results. This is most likely due to missing explicit mathematical descriptions accounting for additional physical phenomena that, at present, are not represented in the current IMR theoretical framework.

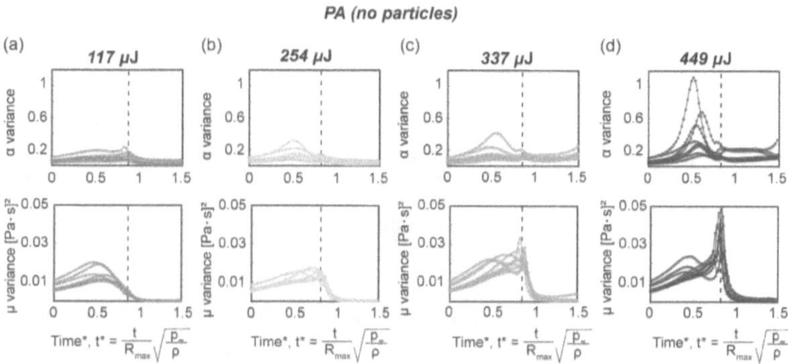

Figure 3.11: Quasi-online IEnKS α and μ estimate variance versus normalized time at nominal laser energies of (a) ∼ 117 μJ (b) ∼ 254 μJ, (c) ∼ 337 μJ, and (d) ∼ 449 μJ. This is for bubble radius time histories in PA with no beads, where the IEnKS was run using an ensemble size of 48. First bubble collapse occurs at approximately the dashed line for all n = 8 samples per experimental condition. Reproduced from Buyukozturk et al. (2022) with permission from Springer Nature, © 2022.

The second representative example we present with the quasi-online (IEnKS) method compares a bubble in a PA hydrogel (no seed particles) at the highest energy to a bubble in a PA hydrogel (paramagnetic seed particles) at the lowest energy. The non-dimensionalized time versus radius curves for each case and online estimation are shown in Figs. 3.12(a,b). These cases have λ_{max} values of 9.5 and 4.7, with low NRMSE values. The data assimilation starts at maximum normalized radius, $R^* = 1$, and normalized time, $t^* = 0$. In Fig. 3.12c, we consider a time-varying α of the PA hydrogel (no seed particles) at highest energy first (λ_{max} = 9.5). There is a gradual increase followed by a decrease in α as the bubble collapses. However, at initial collapse, α experiences a significant increase. Upon the second peak collapse, we observe another increase in the estimation. The time-varying viscosity (Fig. 3.12e) follows a similar trend, where it maintains a steady value until the first collapse, when a large jump increase is observed. Upon second collapse, the estimator adjusts to a higher viscosity estimate again.

We compare this case to a bubble nucleated within PA on a paramagnetic particle at the lowest laser energy ($\lambda_{max} = 4.7$). In this case, α has no discernible jump at collapse, but instead displays a gradual drop-off in the estimation value over time (Fig. 3.12d). Likewise, for the viscosity estimate, we observe relatively constant parameter values in both parameters at collapse, and a slight dip upon first collapse (Fig. 3.12f). It should be noted that while the parameter estimates in subsequent radius peaks are constant, this does not imply a better fit to the model during that portion of the bubble dynamics. Instead, despite using covariance inflation in the quasi-online method, the ensemble eventually converges when the best estimates are found. Thus, subsequent peaks are not a better fit to the model, but rather, the estimator can no longer improve on its estimates for μ and α.

Overall, the low laser energy (paramagnetic seed particle) case exhibits a smoother estimation around first collapse, which is likely due to the smaller exhibited material stretch. This is in stark contrast to the high energy PA (no seed particles) case, with large material stretch, where there is a discernible increase in both parameter estimates and uncertainties at the collapse points. Given the current viscoelastic IMR theoretical framework, the viscosity and strain-stiffening parameters of the surrounding PA hydrogel should be estimated to be the same under both conditions, yet there are observable differences between these cases. These stark contrasts in time-varying material behavior are consistent with previous observations of violent bubble collapses in PA gel. This notion is also reflected by a much higher critical Mach number in the case of no seed particles, compared to that of the paramagnetic particle case (see Fig. 3.10).

The selected examples described in this section illustrate the time-varying material parameter estimates, as well as the model uncertainties with respect to material stretch and laser energy. We see the greatest uncertainty in time-varying material parameters at the bubble collapse points, especially at the first inertial collapse. However, we see a convergence in material parameter variation with decreasing energy for a given case, as well as at lower material maximum stretch regimes, indicating that for lower material stretches or nucleation energies, the DA-IMR framework is appropriate for accurately describing the laser-induced cavitation dynamics.

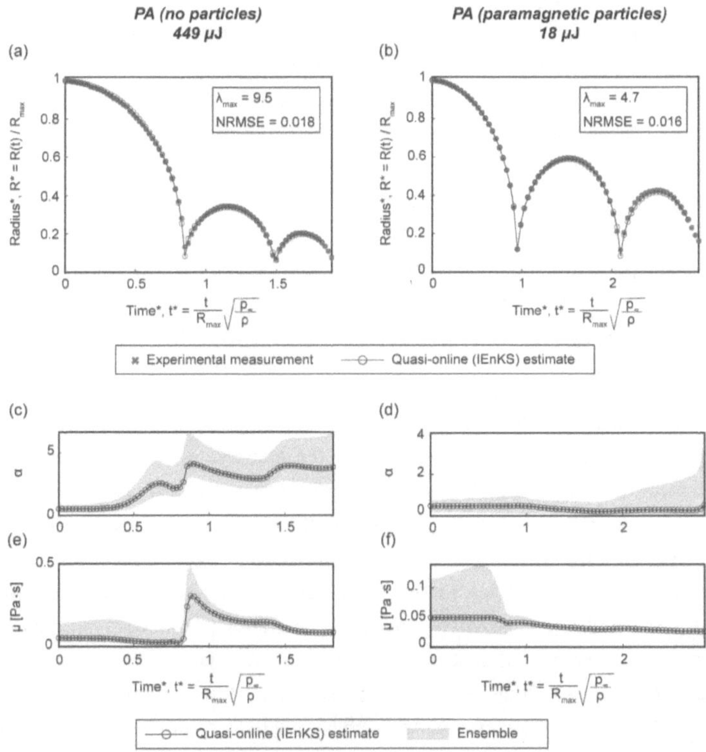

Figure 3.12: Comparison of results of the quasi-online IEnKS fitting for two extreme cases with normalized radius versus time curves in PA (no beads) at nominal laser energy 449 µJ and (b) PA (paramagnetic particles) at nominal laser energy 18 µJ. The (c, d) α (e, f) and μ estimates over normalized time for each case are plotted. Reproduced from Buyukozturk et al. (2022) with permission from Springer Nature, © 2022.

3.2.3 Discussion of Modeling Uncertainties

Overall in section 3.2, we showed that the IEnKS and En4D-Var were able to fit the parameters of a quadratic law Kelvin-Voigt constitutive model to the broad range of data sets, while quantifying associated errors and critical Mach numbers at initial collapse. Important trends were observed given these DA results. With the high temporal resolution available in the experimental data, fitting bubble radius to the first peak only consistently led to a decrease in the NRMSE. While data from previous studies meant that using a long time-scale was vital to obtaining good parameter estimates, it appears that with this image capture rate, limiting to the first peak yields best results. The decrease in result reliability past the first collapse may be a result of the limit on compressibility effects captured in the model. Indeed the Keller-Miksis equation only accounts for compressibility effects near the bubble wall region (Keller and Miksis, 1980; Estrada et al., 2018). While this often suffices, the increase in results NRMSE past the first collapse point indicates that this assumption may not hold, particularly in cases where the critical Mach number exceeds $M_{cr} = 0.08$ (Anderson, 2009). As stretch ratio, and thus Mach number, increases, the physical model used may be limited in its ability to fully capture the bubble dynamics.

It was seen that this effect was more pronounced when increasing the laser energy. While for single-peak fitting, an increase in the stretch ratio did not correlate with an increased fitting error, a higher laser energy consistently increased the NRMSE of results, as seen in figures 3.8, 3.9, and in the increased variance in parameters estimates with the IEnKS with higher energies seen in figure 3.11. This suggests that at higher laser energies, the physics of plasma formation optical breakdown of the polyacrylamide at cavitation inception have an impact beyond our modeling capabilities. The current approach of beginning our data assimilation window at peak bubble radius no longer suffices to avoid capturing this initial breakdown, which affects the physics beyond the initial bubble growth, as assumed by our approach. Lowering energy, for example by introducing seed particles, is thus an effective way to carry out DA-IMR to avoid these effects. However, another way to address this may be to modify the cavitation inception method used, and avoid difficulties caused by the laser nucleation physics altogether. In the next section, we describe our application of the DA-IMR framework to ultrasound-induced cavitation data. There, high intensity focused ultrasound is used to induce cavitation in hydrogels instead of using laser nucleation.

3.3 DA-IMR with Ultrasound-induced Cavitation Data

3.3.1 Experiment and Problem Setup

Mancia, Yang, et al. (2021) introduce 'Acoustic Cavitation Rheometry' as an alternative to the laser-induced cavitation (LIC) rheometry used in sections 3.1 and 3.2. Here, cavitation is generated via focused ultrasound, the negative pressure peaks of which can cause bubble nucleation about existing material defects or microbubbles trapped in a material. This can avoid some of the issues observed in LIC and described in section 3.2.3, while still causing bubbles to grow explosively in media of interest, and attain similar strain rates as with LIC (Wilson et al., 2019). The reader is referred to section 2.1 of Mancia, Yang, et al. (2021) for details on the experimental setup. Key differences with the setup of experiments in section 3.1 and 3.2 are the following. First, agarose gel is used as the cavitation medium rather than polyacrylamide. The 0.3% and 1% gels are prepared following Vlaisavljevich, Lin, Maxwell, et al. (2015), and have quasi-static shear moduli of approximately 380 Pa and 720 Pa, respectively. Next, to induce cavitation, a 16-element transducer array fires 1.5 cycles of focused ultrasound (with a single negative half-period) at 1 MHz with a peak negative amplitude of -24 MPa into the gel. Each bubble, generated by acoustic forcing at the transducer focal point alone (Maxwell, Cain, Hall, et al., 2013), is at least 5 mm away from the previous nucleation site. Images were captured every 2.5 µs (at 400,000 frames per second), giving a better temporal resolution than that at which the DA-IMR methods are validated in Chapter 2.

Given the lack of plasma formation, we now use data from the initial growth of the bubble as well, and are no longer restricted to begin at the first bubble peak. Our data assimilation window thus includes the full first growth and collapse, so while the temporal resolution is worse than that of section 3.2, the number of data points remains large enough to converge on well-fitted material parameters. We continue to use the same bubble dynamics model described in section 2.1, and three distinct material models are tested. First, the Neo-Hookean Kelvin-Voigt model (Gaudron, Warnez, and Johnsen, 2015) used in section 3.1, where the stress integral is given by equation (2.7). Then, two variants of the quadratic law Kelvin-Voigt model (Fung, 2013; Yang, Cramer, and Franck, 2020), with the stress integral given by equation (3.3). The first variant is identical to that used previously in section 3.2, where the shear modulus is fixed at the quasi-static value G_∞. The second generalizes this so that both G and the strain stiffening parameter α can vary. Additionally, instead of using the long-time equilibrium bubble radius to estimate R_0 as done previously, this parameter is here estimated along with the material model parameters, with an initial estimate taken to be the initial bubble radius. Finally, a time-shift parameter t_s is added to our estimation, which is used to initialize bubble growth. This yields a new, expanded state vector for our En4D-Var estimator with up to 5 constant parameters to estimate appended with the generalized QLKV model:

$$x = \{R, \dot{R}, p_b, S, \mathbf{T}, \mathbf{C}, G, \mu, \alpha, R_0, t_s\}. \tag{3.6}$$

For the neo-Hookean Kelvin-Voigt model, $\alpha = 0$ is fixed, for the quasi-static QLKV model we set $G = G_\infty$, and for the generalized QLKV model all parameters are allowed to vary. The main goal of this study is material parameter estimation, so we report results with the En4D-Var method only (as detailed in section 2.2.4). An ensemble size of $q = 48$ is used in all cases, which is found to be sufficient while keeping computational costs low. The initial ensemble is sampled from a Gaussian distribution centered around an initial state vector. The method is implemented with both gels, and for each material model.

3.3.2 En4D–Var Parameter Estimation Results

We now analyze the results obtained with En4D-Var parameter estimation. Initial guesses as to each material property are determined based on the estimates obtained with IMR in table 1 of Mancia, Yang, et al. (2021). For example, in the 0.3% gel case, the initial guesses with all models are $R_0 = 0.5$ μm and $\mu = 0.1$ Pa·s. For the Neo-Hookean model, the shear modulus initial guess is $G = 10$ kPa. For both QLKV models, the initial guess is $\alpha = 0.03$, and while the shear modulus is fixed at $G = 0.38$ kPa in the quasi-static case, the initial guess is $G = 0.5$ kPa in the general QLKV case.

Table 3.3 summarizes the results with En4D-Var for both the 0.3% and 1% agarose specimens. These results are comparable with those obtained with simple least-squares fitting done by Mancia, Yang, et al. (2021) in table 1 for these material parameters. A key difference, however, appears in the comparatively much smaller stress-free radius and stiffening parameter standard deviations. In fact, the estimates for these parameters are close to the initial guess in all cases. This indicates that given these initial guesses, the En4D-Var is unable to improve on these results and converges quickly to the original value. This behavior is further discussed below.

Model	G [kPa]	α [10^{-2}]	μ [Pa s]	R_0 [µm]
0.3% gel				
NH	9.66 ± 0.55	0	0.086 ± 0.028	0.51 ± 0.07
QS QLKV	0.38 ± 0.16	3.0 ± 0.1	0.097 ± 0.026	0.50 ± 0.03
Gen QLKV	0.50 ± 0.05	3.0 ± 0.3	0.094 ± 0.026	0.51 ± 0.06
1% gel				
NH	36 ± 3.57	0	0.14 ± 0.031	1.03 ± 0.04
QS QLKV	7.2 ± 0.33	2.4 ± 0.15	0.16 ± 0.038	1.29 ± 0.06
Gen QLKV	7.7 ± 0.89	2.5 ± 0.12	0.16 ± 0.036	1.29 ± 0.05

Table 3.3: Weighted mean and standard deviation of inferred properties using En4D-Var for 0.3% and 1% agarose specimens. Table reproduced from Mancia, Yang, et al. (2021) with permission from the Royal Society of Chemistry, © 2021.

The radius-normalized RMS errors obtained with the En4D-Var are also similar to those obtained with IMR. For example, in the 0.3% gel case, the normalized RMS errors range from 0.02 to 0.09. with a mean of 0.04 with the Neo-Hookean model. For the quasi-static QLKV model, they range from 0.02 to 0.07 with a mean of 0.05. Finally, for the general QLKV model, the range is 0.02 to 0.06 with a mean of 0.04. The increase in model complexity thus appears to better capture material behavior, reducing the error. Details on these parameter estimates can be found in sections 3 and 4.1-4.3 of Mancia, Yang, et al. (2021), but we here focus on the additional information that our En4D-Var implementation provides.

Indeed, the En4D-Var results can further inform the uncertainties associated with the material property estimates obtained, e.g., due to variations in samples.

Figure 3.13 summarizes the En4D results with the QLKV model for each material property. These histograms combine all final ensemble members across the 19 data sets (with an ensemble size of 48, the total number of ensemble members is thus $19 \times 48 = 912$). An approximately Gaussian distribution is obtained for all four quantities, the mean of which are our estimates for each quantity, thus confirming that the En4D-Var estimates are uniform across all data sets.

Figure 3.14 shows the iterative estimation of each parameter for all 19 data sets, again with the QLKV model. It appears that the viscosity estimates are relatively scattered, which points to a relatively high sensitivity to changes in viscosity in our estimator. By contrast, the spread in stress-free radius, shear modulus and stiffening parameter is quite narrow around the initial guess, indicating that in this regime

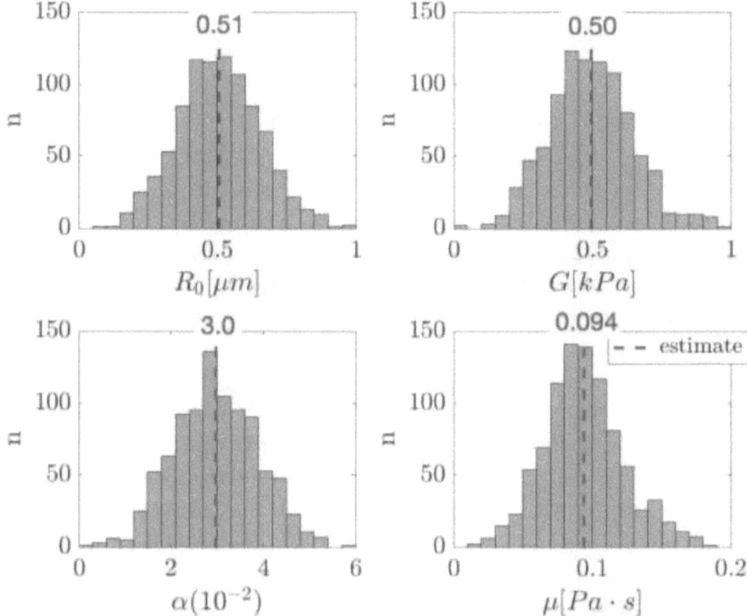

Figure 3.13: Histogram of combined final stress-free radius, shear modulus, stiffening parameter and viscosity ensembles for 0.3% gels with the En4D-var using the QLKV model. Reproduced from Mancia, Yang, et al. (2021) with permission from the Royal Society of Chemistry, © 2021.

and given this guess the estimator could not improve the fit significantly. Overall, this uncertainty assessment demonstrates that the acoustic cavitation extension of IMR applied with data assimilation methods provides robust parameter estimates comparable to those obtained with the traditional IMR optimization approach.

Overall, the En4D-Var method was successfully applied to the ultrasound-cavitation framework for material estimation, or acoustic cavitation rheometry. As compared to the traditional least-squares fitting method, we obtained comparable material parameters in a manner far more scalable and computationally efficient, while additionally providing insight into modeling uncertainties. It has confirmed that acoustic cavitation rheometry is a powerful framework for viscoelastic material property estimation at high strain rates, which avoids some of the issues associated with laser-induced cavitation discussed in section 3.2. Applying En4D to this data

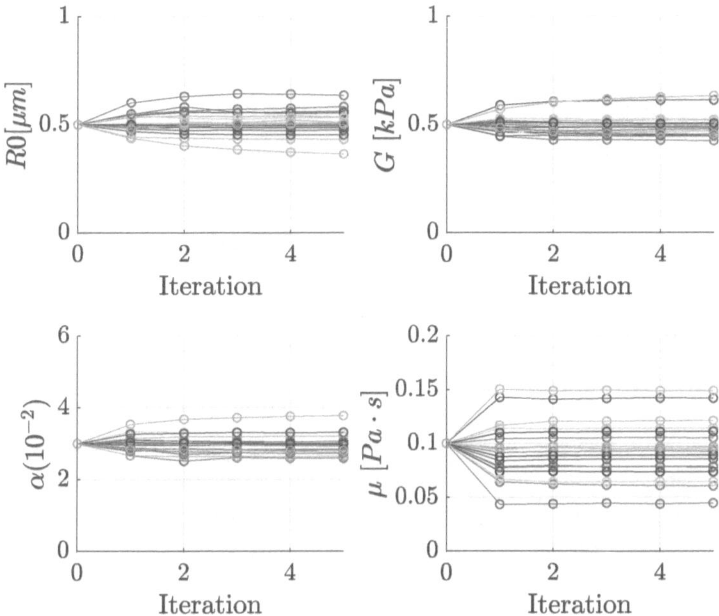

Figure 3.14: **Iterative estimation of the stress-free radius, shear modulus, stiffening parameter and viscosity for 0.3% gels with the En4D-var using the QLKV model. Reproduced from Mancia, Yang, et al. (2021) with permission from the Royal Society of Chemistry, © 2021.**

has shown that DA-IMR is a versatile framework, which in all tested cases has functioned well and provided additional physical insight into the bubble dynamics of the problem.

This concludes the first part of the book, focused on data assimilation for viscoelastic material characterization. While the interaction of ultrasound and bubble dynamics, as examined in section 3.3, play a central role in the following chapters and particularly in Chapter 5, we will now shift to a new application of interest.

Chapter 4

A SINGLE FRAMEWORK FLUID–SOLID NUMERICAL SOLVER FOR BURST-WAVE LITHOTRIPSY SIMULATIONS

Parts of this chapter are adapted from Spratt, Rodriguez, Bryngelson, et al. (2021), Spratt and Colonius (2023) and sections 2.2 and 6.2 of Radhakrishnan et al. (2023). In this chapter, we present our numerical solver implemented for full simulations of burst-wave lithotripsy. In section 4.1, this modeling framework is described, including the 5-equation multiphase model with hypoelasticity, a continuum damage model, a one-way acoustic wave source for the transducer, and the GPU implementation of the code. In section 4.2, our hypoelastic model implementation is validated through comparison to analytical results. In section 4.3, we present first simulations of BWL, and compare simulated stress patterns in stones to experimental photoelastic imaging. Finally, we use the solver in section 4.4 to study the frequency-dependence of stress produced in stones of various sizes and shapes during BWL.

4.1 Numerical Methods

Simulating burst-wave lithotripsy (BWL) requires modeling of various interactions. In the most simple case, we require a solver that can model acoustic (ultrasound) waves traveling through the surrounding fluid and impinging on a solid stone. The elastic response of the kidney stones must be fully represented to determine treatment efficacy and ensure correct interactions with the acoustic field and surrounding fluid. This requires modeling the elastic response of the material in the solid region. Furthermore, the BWL process frequently causes bubbles to cavitate near the proximal surface of the stone. Accounting for this requires a high-order accurate multi-phase compressible flow solver, capable of simulating acoustic wave–bubble interactions, which could include rapid bubble collapse. In all, BWL simulations thus require a multiphase fluid mechanics solver that can additionally model the elastic response of a solid material.

To meet these requirements, we use an interface-capturing method with the 5-equation multiphase flow model (Kapila et al., 2001; Allaire, Clerc, and Kokh,

2002). This is complemented with a hypoelastic model to capture the behavior of solid regions while remaining in an Eulerian framework (Rodriguez and Johnsen, 2019), and a continuum damage model to better understand damage accumulation occurring as a result of the treatment (Cao et al., 2019). Specifically, these models are added to the open-source Multi-component Flow Code (MFC) to carry out our simulations, which uses a diffuse interface method with HLL-type Riemann solvers and supports the 5-equation model (Bryngelson et al., 2021). Spherical acoustic waves generated by the BWL therapy transducer elements are simulated using the volumetric source-term approach of Maeda and Colonius (2017), which can simulate one-way waves off of any smooth surface. Finally, to fully resolve bubbles on the order $O(100\,\mu\text{m})$ in our simulations while capturing the entire stone volume, we require high resolution. This is facilitated by the port of our code to GPUs, without which computational costs for the simulations presented in this paper would have been far greater. We describe each of these models and, in turn, the GPU implementation in the remainder of this section.

4.1.1 The 5-equation Model with Hypoelasticity

We use the hypoelastic material model of Rodriguez and Johnsen (2019), which is based on a small-strain approximation (Eringen, 1962) that facilitates the incorporation of elasticity in an Eulerian framework. This enables adding this model to our existing multiphase flow solver, without sacrificing any of its capabilities. While the cases shown in the present paper only include elastic materials (stones), our implementation uses the Kelvin-Voigt material model, making it generalizable to viscoelastic materials. Details on the Kelvin-Voigt material can be found in Chapters 2 and 3, where it is used to represent the viscoelastic hydrogels characterized using DA-IMR.

The MFC code (Bryngelson et al., 2021) uses the so-called 5-equation multiphase flow model (Kapila et al., 2001; Allaire, Clerc, and Kokh, 2002). *Without* hypoelasticity, it is given by (for 2 components with superscripts $^{(1)}$ and $^{(2)}$):

$$\frac{\partial}{\partial t}\begin{bmatrix} \alpha^{(1)} \\ \alpha^{(1)}\rho^{(1)} \\ \alpha^{(2)}\rho^{(2)} \\ \rho u_i \\ \rho E \end{bmatrix} + \frac{\partial}{\partial x_j}\begin{bmatrix} \alpha^{(1)}u_j \\ \alpha^{(1)}\rho^{(1)}u_j \\ \alpha^{(2)}\rho^{(2)}u_j \\ \rho u_i u_j + p\delta_{ij} - \tau_{ij}^{v} \\ (\rho E + p)u_j - u_i\tau_{ij}^{v} \end{bmatrix} + \begin{bmatrix} -\alpha^{(1)} - K \\ 0 \\ 0 \\ 0 \\ 0 \end{bmatrix}\frac{\partial u_j}{\partial x_j} = \mathbf{0}, \qquad (4.1)$$

with $\alpha^{(k)}$ the volume fraction of the k-th component, and ρ, u, p, E the mixture density, velocity, pressure, and total energy, respectively. τ^v is the viscous stress tensor, and K is the interface compressibility term, which is important in cavitation problems to accurately represent the sound speed in mixture regions (Schmidmayer, Bryngelson, and Colonius, 2020)

$$K = \frac{\rho^{(2)}c^{(2)^2} - \rho^{(1)}c^{(1)^2}}{\frac{\rho^{(2)}c^{(2)^2}}{\alpha^{(2)}} + \frac{\rho^{(1)}c^{(1)^2}}{\alpha^{(1)}}}. \tag{4.2}$$

We now describe modifications of this model to model the elastic response of solids, using a hypoelastic material model (Rodriguez and Johnsen, 2019).

This model relies on using a Lie objective temporal derivative (which preserves thermodynamic consistency) of the stress-strain relation to transform strains to strain-rates (Altmeyer, Rouhaud, et al., 2015; Altmeyer, Panicaud, et al., 2016). Strain rates are computed as gradients of velocity $\dot{\epsilon}_{ij} = \frac{1}{2}\left(\frac{\partial u_i}{\partial x_j} + \frac{\partial u_j}{\partial x_i}\right)$, consistent with fluids. The 5-equation model given by equation (4.1) is modified by first adding an elastic shear stress term $\tau_{ij}^{(e)}$ to the Cauchy stress tensor

$$\sigma_{ij} = -p\delta_{ij} + \tau_{ij}^{(v)} + \tau_{ij}^{(e)}. \tag{4.3}$$

This elastic shear stress will appear in the momentum and energy equations. The total energy is then also modified with an elastic contribution $e^{(e)} = \frac{\tau_{ij}^{(e)}\tau_{ij}^{(e)}}{4\rho G}$, yielding

$$E = e + \frac{\|u\|^2}{2} + e^{(e)}. \tag{4.4}$$

Here, e is the internal energy, and G is the medium shear modulus. Now, the Lie derivative of the elastic stresses is given by

$$\dot{\tau}_{ij}^{(e)} = \frac{\partial \tau_{ij}^{(e)}}{\partial t} + u_k \frac{\partial \tau_{ij}^{(e)}}{\partial x_k} - \tau_{kj}^{(e)} \frac{\partial u_i}{\partial x_k} - \tau_{ik}^{(e)} \frac{\partial u_j}{\partial x_k} - \tau_{ij}^{(e)} \frac{\partial u_k}{\partial x_k}. \tag{4.5}$$

Combining this, and the fact that for an isotropic Kelvin-Voigt material, $\dot{\tau}_{ij}^{(e)} = 2G\dot{\epsilon}_{ij}^{(d)}$ (see section 3.1.1 of Rodriguez and Johnsen (2019)), an evolution equation for the elastic stresses is derived and appended to our modified 5-equation model to yield a final system of equations, given here for two materials:

$$\frac{\partial}{\partial t}\begin{bmatrix} \alpha^{(1)} \\ \alpha^{(1)}\rho^{(1)} \\ \alpha^{(2)}\rho^{(2)} \\ \rho u_i \\ \rho E \\ \rho \tau_{il}^e \end{bmatrix} + \frac{\partial}{\partial x_j}\begin{bmatrix} \alpha^{(1)}u_j \\ \alpha^{(1)}\rho^{(1)}u_j \\ \alpha^{(2)}\rho^{(2)}u_j \\ \rho u_i u_j + p\delta_{ij} - \tau_{ij}^v - \tau_{ij}^e \\ (\rho E + p - \tau_{ij}^e)u_j - u_i\tau_{ij}^v \\ \rho \tau_{il}^e u_j \end{bmatrix} + \begin{bmatrix} -\alpha^{(1)} - K \\ 0 \\ 0 \\ 0 \\ 0 \\ 0 \end{bmatrix}\frac{\partial u_j}{\partial x_j} = \begin{bmatrix} 0 \\ 0 \\ 0 \\ 0 \\ 0 \\ S_{il}^e \end{bmatrix}, \quad (4.6)$$

with

$$S_{il}^e = \rho\left(\tau_{kj}^{(e)}\frac{\partial u_i}{\partial x_k} + \tau_{ik}^{(e)}\frac{\partial u_j}{\partial x_k} - \tau_{ij}^{(e)}\frac{\partial u_k}{\partial x_k} + 2G\dot{\epsilon}_{ij}^{(d)}\right). \quad (4.7)$$

As with the original 5-equation model of MFC, the stiffened-gas equation of state (Le Métayer, Massoni, and Saurel, 2005) closes the system, where for the kth component

$$p_k = (\gamma_k - 1)\rho_k e_k - \gamma_k \pi_{\infty,k}, \quad (4.8)$$

where *gamma* is the specific heat ratio and *pi* is the liquid stiffness, which are fitted to produce correct wave speeds in each material (Le Métayer, Massoni, and Saurel, 2004; Le Métayer, Massoni, and Saurel, 2005).

4.1.2 A Continuum Damage Model

We will use two metrics to determine the efficacy of treatment across our simulations. As brittle materials are expected to fail in tension, we measure the maximum tensile principal stress σ_1 in the stone over the course of a BWL pulse. In 2D, this is given by

$$\sigma_1 = \frac{\sigma_{11} + \sigma_{22}}{2} + \sqrt{\left(\frac{\sigma_{11} - \sigma_{22}}{2}\right)^2 + \tau_{12}^2}, \quad (4.9)$$

where each diagonal stress term σ_{ii} is given by $\sigma_{ii} = -p + \tau_{ii}^{(v)} + \tau_{ii}^{(e)}$.

In 3D, the maximum principal stress is obtained by finding the largest root of the cubic equation:

$$\sigma^3 - I_1\sigma^2 + I_2\sigma - I_3 = 0, \quad (4.10)$$

where the I_i terms are the stress invariants:

$$I_1 = \sigma_{11} + \sigma_{22} + \sigma_{33} \quad (4.11)$$

$$I_2 = \sigma_{11}\sigma_{22} + \sigma_{22}\sigma_{33} + \sigma_{11}\sigma_{33} - \tau_{12}^2 - \tau_{23}^2 - \tau_{13}^2 \quad (4.12)$$

$$I_3 = \sigma_{11}\sigma_{22}\sigma_{33} + 2\tau_{12}\tau_{23}\tau_{13} - \sigma_{11}\tau_{23}^2 - \sigma_{22}\tau_{13}^2 - \sigma_{33}\tau_{12}^2. \quad (4.13)$$

The maximum principal stress will indicate where and in which simulations large peak tensile stresses occur in the stone. This alone is a good predictor of stone damage.

Second, we implement a continuum damage model following Cao et al. (2019). Here, a damage state variable $D(X,t) \in [0,1)$ is introduced, which is calculated as the integral over time

$$D(X,t) = \int_0^t (\overline{\alpha} \max(\sigma_1(X,\tau) - \sigma^*, 0))^s d\tau, \qquad (4.14)$$

where σ_1 is the max. principal stress and σ^*, s, $\overline{\alpha}$ are constant model parameters determined empirically. $D = 0$ indicates a fully intact material, whereas a point in the material where $D = 1$ indicates it is fully damaged. This damage is reflected in the material properties, where in an isotropic linear elastic material, D modifies the local Young's modulus $\mathcal{E}$ as

$$\mathcal{E}(X,t) = \mathcal{E}_0(1 - D(X,t)). \qquad (4.15)$$

In our case, the material stiffness is parametrized by the shear modulus G, and given that $G = \frac{\mathcal{E}}{2(1+v)}$, this yields the following equation for the local time dependent shear modulus

$$G(X,t) = G_0(1 - D(X,t)), \qquad (4.16)$$

where G_0 is the initial material shear modulus. Thus, if D reaches a value of $D = 1$ at a given position in the solid, the material is considered fully damaged, and the shear modulus reduced locally to $G = 0$.

We note that in the results presented in Chapter 5, this damage model is modified and implemented as a one-way model only, meaning that we do not modify local shear modulus as described in equation (4.16). While Cao et al. (2019) looked at shock-wave lithotripsy, our simulations only span a single BWL pulse. Relative to that caused by a shock wave in SWL, the stresses generated by a single BWL pulse with an originally intact stone are small. Indeed, full stone comminution usually requires at least $O(1000)$ ultrasound pulses during BWL. Because of this, the feedback incorporated in the model whereby the damage state modifies local material properties is not necessary for a single BWL pulse, and we only calculate the accumulation of damage D. Thus, the damage model can be thought of as a qualitative way of tracking the accumulation of stress. In Cao et al. (2019), a damage value of $D = 1$ meant that the stone was completely broken, but that is not the case here where damage values are used as a qualitative comparison tool.

The stones used in our simulations in Chapter 5 have identical properties to those of Cao et al. (2019). We thus utilize the same empirical model parameters in equation (4.14), except for σ^* which is reduced to $4\,\mathrm{MPa}$ to account for the time-scale of these simulations of BWL as compared to SWL. Given these modifications, the damage state $D(X,t) \in [0,1)$ serves here as a measure and visualization tool of the accumulation of large maximum principal stress σ_1 over the course of the BWL pulse. Plots of this damage over a cross-section of the stone are used to compare the efficacy of BWL across simulations in Chapter 5.

4.1.3 Spherical Acoustic Wave Generation from a Transducer Source

To model the therapy transducer used in BWL, an acoustic source model capable of generating spherical ultrasound waves is necessary. To this end, we make use of the volumetric source term approach of Maeda and Colonius (2017), which can generate one-way acoustic waves from any smooth surface. In this case, we insert individual spherically focused transducer elements with a given aperture and focal length, each of which serves as a source for a spherically focused sine wave. Following section 4.2 of Maeda and Colonius (2017), source terms are added to the mass, momentum (in the longitudinal and radial direction), and energy equations in our model, given by

$$
\Omega(\xi,t) = \begin{bmatrix} f(t) \\ g_z(\xi,t) \\ g_r(\xi,t) \\ \frac{c_0^2 f(t)}{\gamma-1} \end{bmatrix},
\tag{4.17}
$$

where

$$
f(t) = \frac{p_a}{c_0}\sin(\omega(t-t_0)) + \frac{p_a}{A}\left(\frac{1}{\omega}\cos\left(\omega^2(t-t_0)\right) - \frac{1}{\omega}\right)
\tag{4.18}
$$

$$
g_z(\xi,t) = -p_a\sin(\omega(t-t_0))\cos(\xi)
\tag{4.19}
$$

$$
g_r(\xi,t) = -p_a\sin(\omega(t-t_0))\sin(\xi).
\tag{4.20}
$$

Here, f is the ultrasound frequency, p_a is the pressure pulse amplitude, c_0 the speed of sound, $\omega = 2\pi f$ the angular frequency, and A the element aperture. ξ is the polar angle parametrizing the transducer element as $(x,y) = [r_0\cos(\xi), r_0\sin(\xi)]$. The full model derivation and examples of its use are given in Maeda and Colonius (2017).

For the simulations of Chapter 5, we implement a custom transducer surface with 18 spherically focused elements placed in two concentric rings (of 6 and 12 elements). The transducer element pressure amplitudes are calibrated to match experimental waveforms. Details are provided in the description of the problem setup in section 5.1, and in section 5.2 where a virtual array is created to simulate a transducer within a reduced domain size.

4.1.4 GPU Acceleration

For the past two decades, CPU clock speed increase has slowed and come to a plateau. Instead, much of the increase in computational power has come from increased parallelization, particularly through the use of GPUs. Modern supercomputers make effective use of GPU acceleration, which has become central to the continued scaling of computational capabilities. Given CFL number limitations that govern simulations in computational fluid dynamics, the minimization of wall time for each time step is critical. To this end, the MFC code, including all models described thus far in section 4.1, has been ported to GPUs. Details of this implementation can be found in Radhakrishnan et al. (2023), and are summarized here.

The GPU implementation is done using OpenACC (Wienke et al., 2012), which is a directive-based programming language. OpenACC directives are read at compile-time, only when compiling with GPU-enabled compilers. Thus, GPU kernels are only created when necessary hardware is available, and OpenACC directives are otherwise ignored. This enables the use of a single codebase for CPU and GPU execution, while retaining high performance comparable to using GPU-specific programming languages such as CUDA (Khalilov and Timoveev, 2021). Furthermore, OpenACC is not GPU-vendor dependent, and has increased support across various computing ecosystems, meaning a flexible and future-proof implementation as GPU architecture evolves (Jarmusch et al., 2022). Despite its relative ease of implementation, it enables custom parallelization strategies and data management between CPU host and GPU devices, serving the needs of a relatively large Fortran code such as MFC.

The results are a performant and highly scalable GPU-accelerated code. Speedup numbers vary depending on the hardware and models used. In one representative test case reported in Radhakrishnan et al. (2023), we see a 300× speedup when comparing a single NVIDIA V100 GPU to a POWER9 CPU core. In another test case comparing a full GPU node (4 NVIDIA V100 GPUs) to a CPU node (128 cores AMD EPYC 7742) on the Expanse supercomputer at the San Diego Supercomputer Center, a 5.8× speedup is achieved with the hypoelastic and damage models turned on in a node-to-node comparison (see table 4.1 and Appendix B for full results of this example). We note that this speedup would be even more significant if using NVIDIA A100 GPUs, which are approximately 1.7 times faster than NVIDIA V100s.

The code sees near-ideal weak scaling (where the problem size scales with the number of processors) up to $O(10000)$ GPUs. That is, running a test 3D two-component case with 1 million points/GPU is only 3% slower on 13824 GPUs than on a single GPU. Strong scaling (where the problem size is fixed for each run) sees MFC retain 84% of ideal performance when increasing the number of GPUs 8-fold, from 8 to 64 GPUs. This decreased performance is attributed to the increased MPI communication required as the number of GPUs increases. We note that for all GPU simulations presented in Chapter 5, the number of GPUs did not exceed 32, where scaling performance remains near-ideal (see figure 2 in Radhakrishnan et al. (2023)). Overall, the GPU implementation of the code enables runs at much higher resolutions than previously possible with the CPU code alone.

When the hypoelastic model is activated, we have seen in section 4.1.1 that up to 6 additional equations must be solved. That is, there are up to 6 distinct elastic stresses σ_{ij} in 3D, as this is a symmetric 3×3 tensor. Also, calculating the damage state following the method described in section 4.1.2 requires calculating the maximum principal stress at each point in the domain. These models thus both increase the computational cost of simulations. Table 4.1 compares run time per time-step for a 3D test case with the base version of MFC, with the hypoelastic model turned on, and with both the hypoelastic and damage models. This shows that the GPU

Version	Base MFC	Hypoelastic	Hypoelastic and Damage
CPU node (128 cores)	8.8 s	19.4 s	22.5 s
GPU node (4 GPUs)	2.0 s	3.7 s	3.9 s

Table 4.1: Total time per time-step comparison for a 3D MFC test run with 10 million grid points with two components, with different models activated.

code reduces the additional burden of the implemented models. On one CPU node, the added cost of turning on both models is 155%. By contrast, this added cost is reduced to 98% on one GPU node. The added cost of including just the hypoelastic model is similarly lower (84% vs 119%) on GPUs. This is likely because the performance of the GPU code is more limited by routines with intense memory transfer requirements. Calculations required for both the hypoelastic and damage models are highly parallelizable, and thus constitute a smaller added cost relative to the entire time step operations on GPUs, as detailed in Appendix B.

Having described our numerical framework including implemented hypoelastic and damage models, acoustic source model, and the GPU acceleration of the code, we now turn to validation cases, particularly for the implemented hypoelastic model.

4.2 Model Validation

4.2.1 Comparison to 1D Analytical Solution

In 1D, following Gavrilyuk, Favrie, and Saurel (2008), we can obtain analytical solutions to Riemann problems in elastic media as long as ρG is kept constant. We consider two such 1D Riemann problems in elastic media taken from Gavrilyuk, Favrie, and Saurel (2008) and Rodriguez and Johnsen (2019), one of which was also reported in Spratt, Rodriguez, Bryngelson, et al. (2021). The first is an impact problem in an elastic material with shear modulus $G = 500\,\mathrm{MPa}$. There is a velocity discontinuity at the interface, and the full initial conditions are given by

$$(\rho, u, p, \tau_e) = \begin{cases} (1000, 10, 1 \times 10^5, 0) & \text{for } x \in [0, 0.5], \\ (1000, -10, 1 \times 10^5, 0) & \text{for } x \in [0.5, 1]. \end{cases} \tag{4.21}$$

The resulting density, velocity, pressure, and normal stress at $t = 64\,\mu\mathrm{s}$ are shown in figure 4.1. The simulation is run on a 1D grid with 400 points. We see good agreement between the simulation results and analytical solution.

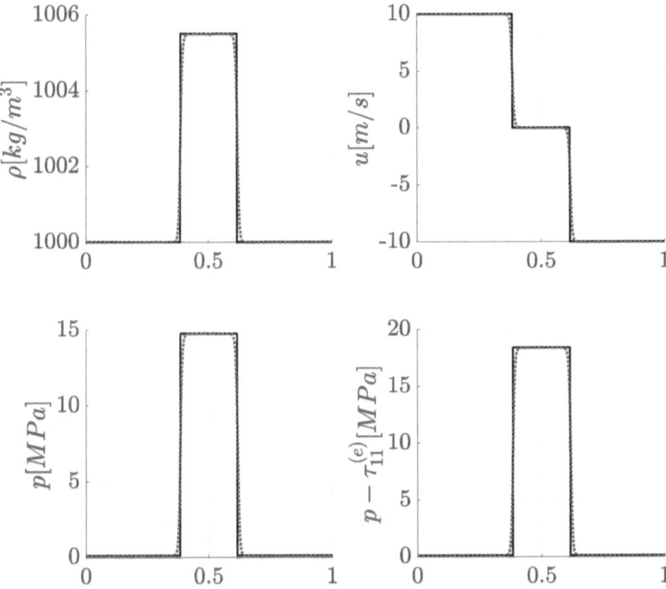

Figure 4.1: Comparing the simulated results (blue dotted line) to the exact solution (black line) for a 1D impact problem with initial conditions described by (4.21) **at** $t = 64\,\mu s$.

The next validation case is akin to a Sod shock tube with a $1000 : 1$ pressure ratio, but with an added velocity discontinuity in the y-direction at the interface in an elastic material with high shear modulus $G = 10\,\text{GPa}$. The initial conditions are

$$
(\rho, u, v, p, \tau_e) =
\begin{cases}
(10^3, 0,\ \ 100, 10^8, 0) & \text{for } x \in [0, 0.5), \\
(10^3, 0, -100, 10^5, 0) & \text{for } x \in [0.5, 1].
\end{cases}
\tag{4.22}
$$

To simulate this case with MFC, we run a 'quasi-1D' simulation with the 2D code. By this, we mean that initial conditions are uniform along the y-direction (which only has a few grid points). This way, we achieve the y-velocity discontinuity required in this example. We plot the resulting state (along the centerline) at time $t = 64\,\mu s$ in figure 4.2. Density, velocity in the x-direction, pressure, normal stress, velocity in the y-direction, and elastic shear stress τ_{12} are shown for $x \in [0, 1]$. Note

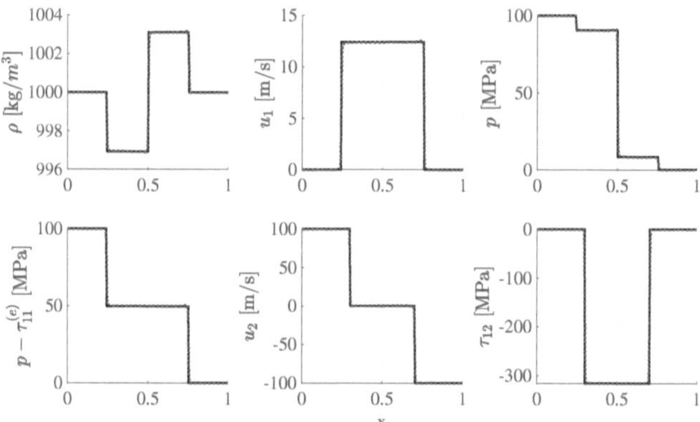

Figure 4.2: Comparing the simulated results (blue dotted line) to the exact solution (black line) for the 1D Riemann problem described by initial conditions (4.22) at $t = 64\,\mu\text{s}$.

that these values are plotted along the $y = 0$ axis, but are equal for any y as the initial condition is uniform in the y-direction. Five waves are observed: a shock propagating to the right, a rarefaction propagating to the left, a contact discontinuity in the center, and shear waves propagating in both directions. Simulation results show good agreement with the exact solution, notably demonstrating correct p-wave and s-wave speeds in this elastic material.

4.2.2 2D Bi-Layered Media Simulations

Next, we consider a two-dimensional simulation of bi-layered media to examine transmitted s- and p-waves through a water–elastic-medium interface. This wave transmission is vital to the understanding of the elastic response of stones to BWL. Here, an acoustic source placed in the top fluid layer, 500 m above the interface, generates one cycle of a 10 Hz Ricker wave. This waveform, which is a derivative of a Gaussian pulse, is frequently used as a surrogate for seismograph data (Ryan, 1994). The elastic material has density $\rho = 2500\,\text{kg/m}^3$ and theoretical wave-speeds $c_L = 3400\,\text{m/s}$ and $c_T = 1963\,\text{m/s}$. A probe is placed 500 m below the material interface, and 2000 m from the source in the x-direction. The full setup is shown in figure 4.3.

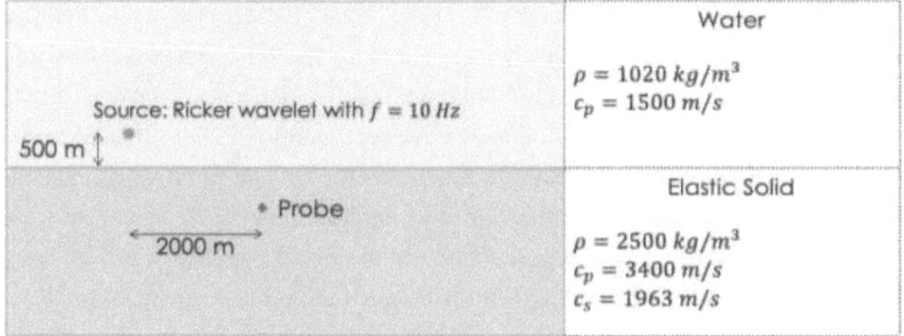

Figure 4.3: Sketch of the setup for a 2D bi-layered media problem, showing the location of wave source and probe as well as material properties of each layer.

This example is taken from Komatitsch, Barnes, and Tromp (2000), and thus we can compare the resulting wave pattern to that observed in their spectral-element code simulations. Figure 4.4 shows the velocity magnitudes in the domain at two distinct times, where the reflected and transmitted waves can be observed.

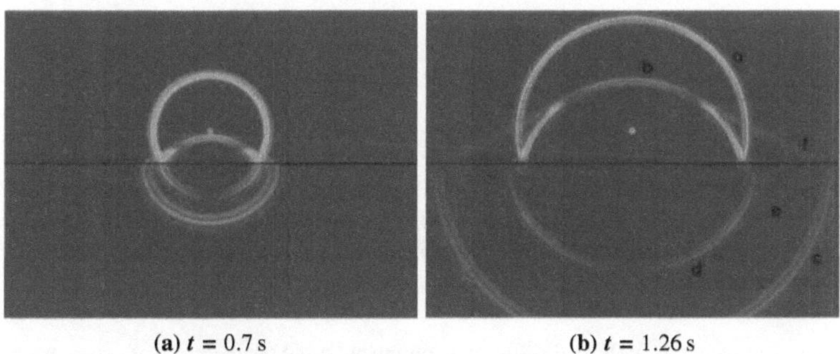

(a) $t = 0.7$ s   (b) $t = 1.26$ s

Figure 4.4: Simulated velocity magnitudes in bi-layered media. A source in the liquid (top half) emits a Ricker wavelet which reflects and propagates through the material interface with the elastic solid (bottom half). Resulting wave patterns are shown at two different times t as labeled.

The simulation displays all expected waves. These are labeled as follows: The initial wave emitted from the source (a), a reflected (b) and transmitted (c) p-wave, a transmitted s-wave (d), and two refracted waves (e,f). Calculated p and s-wave speeds in the elastic medium match the expected theoretical values within 0.1%, and comparing to figure 1 of Komatitsch, Barnes, and Tromp (2000) shows great alignment.

We also compare our simulation results at the probed location (shown as the blue diamond in figure 4.3) to an analytical solution. For this comparison, we use analytical results obtained with the Gar6more2D solver of Diaz and Ezziani (2010). They compute the analytical solution to wave propagation in 2D bi-layered media using the Cagniard-De Hoop method (Cagniard et al., 1963; De Hoop, 1960). Details of their implementation can be found in Diaz and Ezziani (2010). Specifically, we plot the velocity in the x and y-direction for the first 1.5 s from wave emission in figure 4.5 This shows good agreement between the probed velocities from our

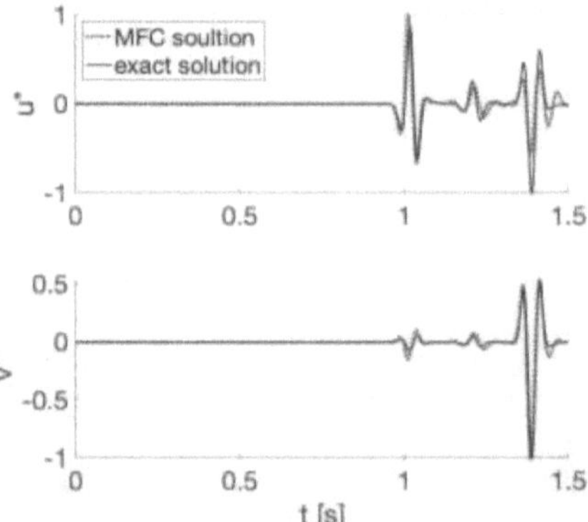

Figure 4.5: Velocity in the x (top) and y-direction (bottom) at the probe point in the elastic solid layer, located 1000 m below the source in the y-direction and 2000 m past the source in the x-direction. The probe is 500 m below the water-solid interface.

simulation and the analytical solution, showing accurate propagation of elastic waves through the interface and solid material. Three waves are visible in this plot. The first, at $t \approx 1$ s, corresponds to the transmitted p-wave labeled (c) in figure 4.4. The second, at $t \approx 1.2$ s is the refracted wave labeled (e), and the final wave at $t \approx 1.4$ s is the transmitted s-wave labeled (d).

The three validation cases presented in section 4.2 have shown the accuracy of our simulation framework, as compared to analytical results. Section 4.3 presents a validation case for a simulation of BWL, where results are compared to experimental stress measurements.

4.3 Comparison of BWL Simulations to Experimental Photoelastic Stress Images

Experimentally, it can be difficult to determine the optimal transducer setup for stone comminution. Surrogate kidney stones can be used, and time to comminution recorded in various settings, but this does not help understand elastic wave propagation patterns leading to stone damage. To address this, Sapozhnikov, Maxwell, and Bailey (2020) used an ingenious method to visualize stresses occurring in stones in real-time during experiments. For this, they make use of photoelastic stress imaging. This method works by shining a polarized light through a stone (which must be transparent, and optically isotropic) undergoing BWL. The local stresses in the stone modifies the polarization of the light shining through it, and thus the projected light displays patterns of stress inside the stone. However, this cannot be done in retrieved human kidney stones, or other typical kidney stone phantoms used experimentally, as the light beam cannot traverse these. Instead, Sapozhnikov, Maxwell, and Bailey (2020) use transparent, optically isotropic epoxy, which is prepared to match kidney stone sound speeds and density relatively well. Another advantage of these experiments, is that they are a good point of comparison for BWL simulations.

In this section, we show simulations of BWL using the numerical framework described in section 4.1, and compare them to the photoelastic images obtained by Sapozhnikov, Maxwell, and Bailey (2020) and Maxwell, MacConaghy, et al. (2020). We replicate their experimental setup and use cylindrical stones with length 20.8 mm, diameter 6.3 mm. The stone material properties are identical to the epoxy stones used: $\rho = 1100\,\text{kg/m}^3$, $c_L = 2440\,\text{m/s}$, $c_T = 1295\,\text{m/s}$. Figure 4.6 shows the setup of this simulation, where the transducer is simplified as a single spherically focused element with aperture 90 mm and focal length 70 mm, and a cylindrical stone is submerged in water. The transducer fires 20 cycles of ultrasound at 340 kHz.

In figure 4.7, we show the pressure in the surrounding liquid, and maximum principal stress σ_1 in the stone during the BWL pulse at three distinct times.

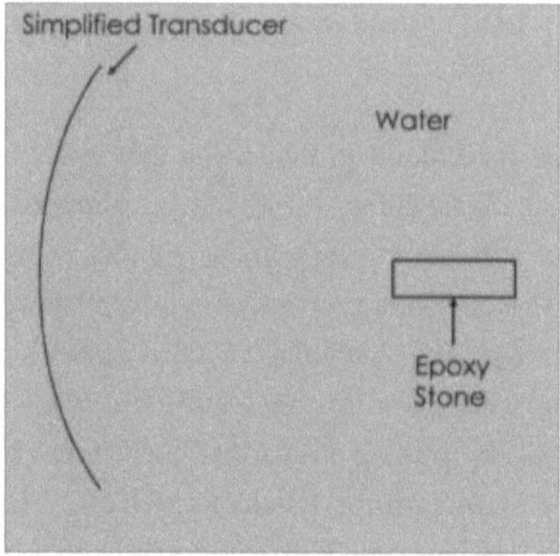

Figure 4.6: Setup of the simulation with a submerged epoxy stone on the right and simplified, single-element ultrasound transducer on the left, both submerged in water.

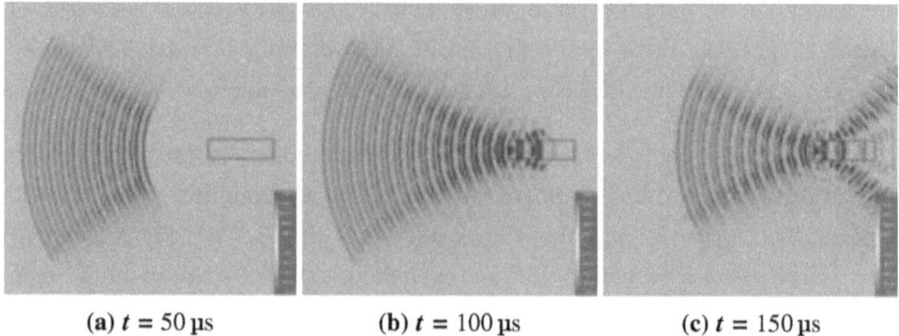

(a) $t = 50\,\mu s$ (b) $t = 100\,\mu s$ (c) $t = 150\,\mu s$

Figure 4.7: Propagation of pressure in the fluid and maximum principal stress in an epoxy stone submerged in water as a result of a 20-cycle BWL ultrasound pulse from a single-element transducer at 340 kHz, shown for three distinctive times as labeled.

We compare this to experimental photoelastic images. Using the method outlined by Sapozhnikov, Maxwell, and Bailey (2020), computational photoelastic images can be obtained from the stresses calculated in our simulations. Section 2 of Sapozhnikov, Maxwell, and Bailey (2020) derives light propagation equations dependent on local stresses. These equations are integrated over the stone cross-section to obtain a 2D distribution of light intensity, matching what would be seen in experimental photoelastic images. Details on this method can be found in sections 2 and 3 of Sapozhnikov, Maxwell, and Bailey (2020). In figure 4.8, we compare the computational photoelastic images obtained from the MFC simulation (top) to experimental images of Maxwell, MacConaghy, et al. (2020) and Sapozhnikov, Maxwell, and Bailey (2020) (bottom).

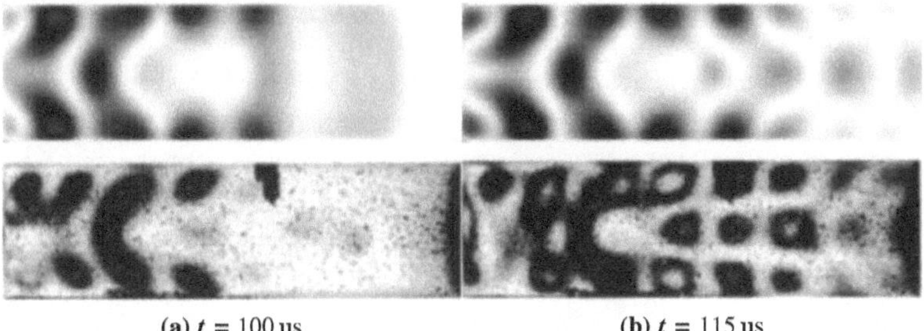

(a) $t = 100\,\mu s$ (b) $t = 115\,\mu s$

Figure 4.8: Comparison of computational photoelastic stress images in an epoxy stone obtained in our simulation (top) and experimental images obtained by Sapozhnikov, Maxwell, and Bailey (2020) which are also reported in Maxwell, MacConaghy, et al. (2020) The bottom experimental images are reprinted with permission from Sapozhnikov, Maxwell, and Bailey (2020). © 2020, Acoustical Society of America.

While quantitative comparisons with photoelastic images are challenging given the dependence of computational photoelastic modeling on empirical parameters (Sapozhnikov, Maxwell, and Bailey, 2020), good qualitative agreement between our simulation results and the experimental data is observed. This shows that our simulation framework can adequately model the behavior of elastic waves in a stone exposed to a BWL pulse. We now use the hypoelastic model to study the frequency dependence of stress patterns in stones during BWL.

4.4 Frequency–Dependence of Stresses Generated by BWL

Having validated our numerical framework, including with a BWL example, we apply it to simulate BWL at various frequencies, and in stones with varying shapes. The goal is here to determine a frequency which optimally damages stones of typical composition and size seen in human kidney stones. In Chapter 5, the problem is complicated by introducing bubbles to shield the stone, but we begin here by applying our framework in cases without bubbles present. Thus, we aim to determine a *baseline* frequency to be used going forward.

Experimental studies have explored BWL treatment frequency (Maxwell, Cunitz, et al., 2015; Maxwell, MacConaghy, et al., 2020). These have led to the generally accepted range of 350 kHz to 400 kHz, which is used in clinical trials (Harper, Metzler, et al., 2021; Harper, Lingeman, et al., 2022). By deriving an analytical model for mechanical stress in a spherical stone during BWL, Sapozhnikov, Maxwell, and Bailey (2021) explore the stone-size–dependence of maximum stress at a given frequency (or conversely, the frequency–dependence for a fixed stone size). Their study focuses on smaller stones, and shows that higher frequencies of ultrasound plane-waves increase stress in small stones. With plane waves, they show that stresses in circular stones remain approximately equal to the pressure wave magnitude when the wavelength is much larger than the stone diameter. However, as the frequency increases and the wavelength drops below twice the stone diameter, stress in the stone can increase to many times this value. Here, we calculate stresses generated in circular and rectangular stones at various ultrasound frequencies guided by their findings, but using our numerical framework with a spherical transducer element, and stone composition matching what will be used for full BWL simulations in Chapter 5.

Given the large beam-width (compared to the stone diameter) in BWL, we expect our results with the spherical waves to closely match the observations of Sapozhnikov, Maxwell, and Bailey (2021). They find that for calcium oxalate monohydrate (COM) stones with $\rho = 1823 \, \text{kg/m}^3$ $c_L = 4476 \, \text{m/s}$ $c_T = 2247 \, \text{m/s}$, large stresses are generated for ultrasound frequencies around 350 kHz for 5mm diameter stones, and 175 kHz for 10mm stones. We note that higher stress peaks are observed at higher frequencies, but with narrow frequency bands. In practice, using the listed frequencies where high stresses are sustained over a larger frequency range is preferable. In a setup matching that of figure 4.6, we expose circular stones with a 5mm and 10mm diameter to three ultrasound frequencies ranging from 175 kHz to 700 kHz. Table 4.2 reports values for $\max(\sigma_1)/p_0$, the maximum principal stress over the simulation normalized by the focal pressure, which is calibrated to $p_0 = 6.5 \, \text{MPa}$ for all tested frequencies.

	frequency [kHz]	$\max(\sigma_1)/p_0$
5 mm stone	175	2.18
	350	**3.22**
	700	2.17
10 mm stone	**175**	**3.18**
	350	1.97
	700	2.80

Table 4.2: Maximum value of the normalized maximum principal stress over a BWL simulation with a single spherical element at three fixed frequencies, for circular stones with diameter 5 mm and 10 mm; bold values represent the maximum stress values in each case.

The trend observed by Sapozhnikov, Maxwell, and Bailey (2021) is also seen in our results, with the largest stresses observed at 350 kHz for the smaller stone, and 175 kHz for the larger stone. $\max(\sigma_1)$ fields in the stone for each of these cases are shown in figure 4.9 for the 5 mm-diameter stone, and figure 4.10 for the 10 mm-diameter stone.

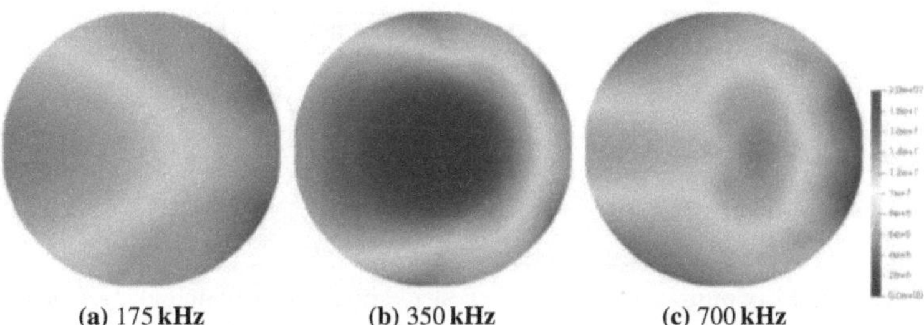

| (a) 175 kHz | (b) 350 kHz | (c) 700 kHz |

Figure 4.9: Maximum value of σ_1 in a 5 mm diameter circular stone over the course of a BWL simulation with a single spherical transducer element at three test frequencies.

Large areas of high tensile stress are observed at 350 kHz for the 5 mm stone, and at 175 kHz for the 10 mm stone, as expected. We note the interesting pattern observed for the 10 mm stone at 700 kHz in figure 4.10c. This case, which table 4.2 also showed has a relatively large $\max(\sigma_1)$ value, corresponds to one of the narrower stress peaks discussed in Sapozhnikov, Maxwell, and Bailey (2021). At this frequency, resonance in the stone is causing localized stress peaks of high magnitude. However, in practice, this regime is difficult to reach unless stone size

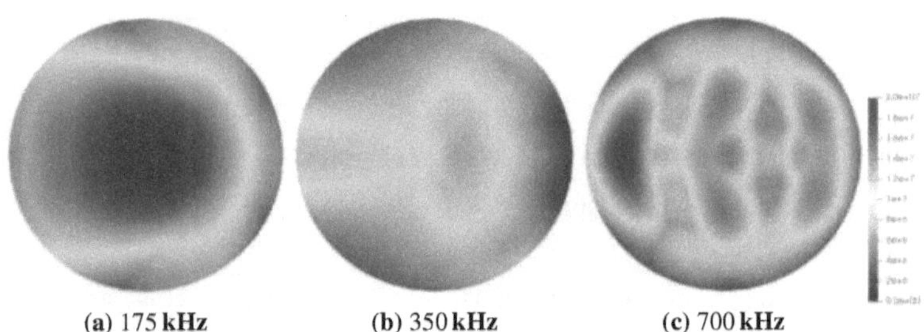

(a) 175 kHz   (b) 350 kHz   (c) 700 kHz

Figure 4.10: Maximum value of σ_1 in a 10 mm diameter circular stone over the course of a BWL simulation with a single spherical transducer element at three test frequencies.

is precisely known a priori. Furthermore, while the peak stress values in this case are high, they are far more localized.

To examine the shape-dependence of these results, we calculated the equivalent $\max(\sigma_1)$ values in rectangular stones, which are shown in table 4.3

	frequency [kHz]	$\max(\sigma_1)/p_0$
	175	2.35
5×5 mm	**350**	**2.38**
	700	1.77
	175	**4.47**
10×5 mm	350	1.85
	700	1.54

Table 4.3: Maximum value of the normalized maximum principal stress over a BWL simulation with a single spherical element at three fixed frequencies, for rectangular stones with width 5 mm, and length 5 mm and 10 mm, respectively; bold values represent the maximum stress values in each case.

The same trends are observed, with highest values of $\max(\sigma_1)$ at 350 kHz in the small stone, and 175 kHz in the large stone. Representative stress patterns are shown for the 5 mm length stone in figure 4.11.

While the $\max(\sigma_1)/p_0$ values reported for this rectangular stone in table 4.3 were similar for 175 kHz and 350 kHz, figure 4.11 shows a preferable stress pattern with the higher frequency of 350 kHz, where the peak stresses occur across a large region of the stone, centered in the middle of the stone.

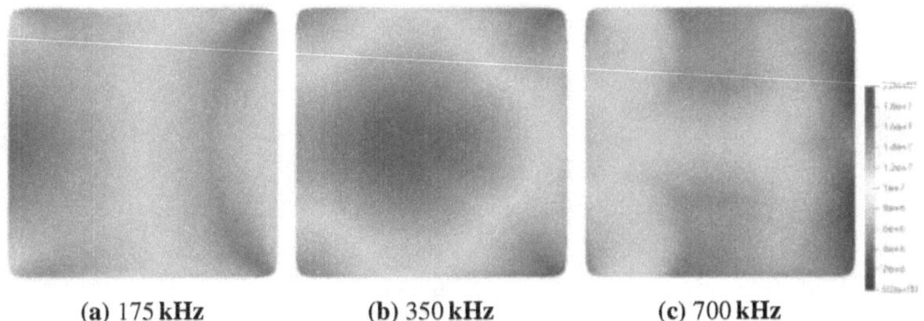

(a) 175 kHz (b) 350 kHz (c) 700 kHz

Figure 4.11: Maximum value of σ_1 in a 5 mm length rectangular stone over the course of a BWL simulation with a single spherical transducer element at three test frequencies.

Overall, these results have shown that previous experimental results, as well as theoretical results with plane waves, appear to hold for BWL with spherically focused ultrasound waves. Moreover, these results hold for different stone shapes, with length-scale seemingly the most important factor to predict frequencies causing larger peak stresses. While we see high peak stresses at higher resonant frequencies in certain cases, as seen with the 10mm diameter stone at 700 kHz, a frequency near 350 kHz is most reliable, as Sapozhnikov, Maxwell, and Bailey (2021) showed that high stresses are generated in a broader band of frequencies around this value. In subsequent simulations in Chapter 5, we use 375 kHz as our baseline frequency, which is near this 350 kHz value, but appeared in testing to yield slightly higher stresses in the material used, which has higher wave-speeds. Given that higher frequencies yield higher stresses in smaller stones, testing at a slightly higher frequency is also advantageous as it will continue to deliver high stress to the stone as it fragments and its length scale is reduced.

Chapter 5

NUMERICAL INVESTIGATION OF BUBBLE DYNAMICS AND DAMAGE IN BURST-WAVE LITHOTRIPSY

Parts of this chapter are adapted from Spratt and Colonius (2023). In Chapter 4, we presented BWL simulation results involving the interaction of focused ultrasound and stones. However, as discussed in section 1.6 of Chapter 1, it has been observed that bubbles can nucleate during lithotripsy and impact treatment efficacy (Maeda, Kreider, et al., 2015; Maeda, Maxwell, et al., 2018; Maeda, Colonius, et al., 2018; Hunter et al., 2018). In particular, Maeda, Maxwell, et al. (2018) looked at energy shielding of bubble clouds over single pulses and saw that they can reduce delivery of acoustic energy to the stone by as much as 90%. In this chapter, using the numerical setup described in detail in Chapter 4, we quantify the impact of shielding on wave propagation and the resulting damage within the stone. Additionally, we investigate further strategies to minimize shielding by modulating the ultrasound frequency of individual transducer elements, introducing a secondary low-frequency ultrasound wave that causes bubbles to collapse ahead of the stone. We examine different configurations of this strategy, aiming to maximize the efficiency and efficacy of treatment in the presence of bubbles.

As an outline of the chapter, we begin by describing the problem setup in section 5.1. Then, the creation of a virtual transducer array to replicate waves generated from a transducer with a much smaller acoustic source to reduce computational costs is detailed in section 5.2. In section 5.3, we use this framework to show the impact that the presence of bubbles near the proximal part of the stone has on the stresses and damage generated in the stones. Section 5.4 examines bubble dynamics when modulating transducer element frequencies according to two distinct strategies, to attempt to encourage bubble collapse. In section 5.5, we run full (wave–bubble–stone) BWL simulations once again with these shielding mitigation strategies to test their efficacy. Results are discussed in section 5.6.

5.1   Problem Setup

A simple sketch of a side-view of the experimental setup for BWL which our simulations replicate is shown in figure 5.1. We use a spherically focused transducer with focal length $F = 150\,mm$ and total aperture is $A = 180\,mm$. The transducer is focused on the center of a cylindrical stone with length 6 mm and diameter 5 mm. The stone (and transducer array) are submerged in water. The sketch is not drawn to scale, as the transducer aperture and focal length are much larger than the stone length scale. In section 5.2, we provide more details on the transducer, and show how a virtual array is created to replicate it with a much smaller aperture and focal length. This way, we minimize computational costs by reducing the required domain size.

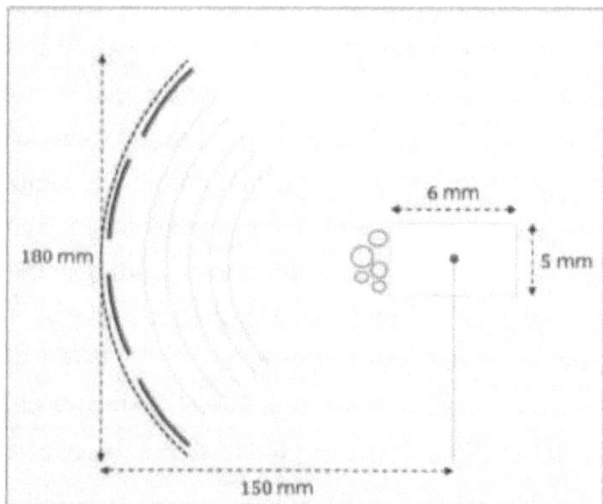

Figure 5.1: Simplified sketch of the setup used for BWL simulations throughout Chapter 5.

The kidney stones in our simulations are modeled using the material parameters of a BegoStone (Liu and Zhong, 2002), an artificial surrogate for kidney stones, which has been used in various experimental studies of SWL and BWL (Maxwell, Cunitz, et al., 2015; Zwaschka et al., 2018). The properties of BegoStone fall within the range of density and sound speeds seen in human kidney stones. In our simulations, we use a stone density of $\rho = 1995kg/m^3$, and longitudinal and transverse wave speeds $c_L = 4159m/s, c_T = 2319m/s$ to match that of Cao et al. (2019) for a BegoStone prepared as detailed by Esch et al. (2010). The cylindrical stone has a length of 6 mm and a diameter of 5 mm, and is submerged in water.

Finally, to model a bubble cloud in front of the stone, we randomly distribute air bubbles with radii following a log-normal distribution with a mean of 250 μm. The standard deviation is determined through observation of experimental and simulated bubble clouds in Maeda, Maxwell, et al. (2018). Our setup is agnostic as to how bubbles formed, and we begin simulations with air bubbles already present. Given the small time-scale of simulations, bubbles will only collapse to a certain point. While bubbles nucleated during BWL may contain some water vapor, the dynamics of air bubbles will adequately capture these dynamics up to final collapse.

Maeda, Maxwell, et al. (2018) performed a number of simulations with varied initial bubble cloud thickness h and void fraction β_0. Using a subgrid bubble model, they excite bubble nuclei with ultrasound analogous to that used in BWL, and track the evolution of these bubble clouds via their total area A when projected on a 2D plane. They observe that the bubble cloud projected area plateaus near a value dependent on the initial thickness and void fraction of the bubble cloud. Furthermore, this projected area value correlates to shielding factor, with larger values of A corresponding to higher shielding. Given the use of resolved bubbles in our simulations, we do not simulate the growth phase of bubbles from the original nuclei, and instead initialize our bubble clouds with a given projected area corresponding to these plateau values. The projected area value $A = 2\,\text{mm}^2$ used in the following sections corresponds to a case with initial void fraction of $\beta_0 = 8 \times 10^{-5}$. A denser bubble cloud with projected area $A = 3\,\text{mm}^2$ is also considered. Given the results of Maeda, Maxwell, et al. (2018), both bubble clouds are expected to shield the stone effectively.

We set the mean bubble radius of 250 μm to match the upper bound for radii values seen in experiments and simulations of BWL with bubbles present (Maeda, Maxwell, et al., 2018; Maeda, Colonius, et al., 2018), while remaining large enough to be resolved in our simulations. This enables a resolution of ~ 50 grid points per bubble diameter. Furthermore, while small bubbles readily oscillate when exposed to ultrasound at treatment frequencies, the dynamics of these larger bubbles are less sensitive to these waves, as will be shown in simulations in section 5.4. The persistence of larger bubbles over multiple ultrasound pulses is a likely cause for reduced treatment efficacy, which our setup can examine. Standoff distance of the bubble cloud is chosen to match observed values in Maeda, Maxwell, et al. (2018). A representative setup for a full simulation with a bubble cloud present is shown in the next section, in figure 5.4b.

5.2 Virtual Transducer Array Calibration

To accurately capture the focused ultrasound waves used experimentally (Maeda and Colonius, 2017; Maeda, Maxwell, et al., 2018), we model an 18-element transducer, which consists of an inner ring of 6 elements, and an outer ring of 12 elements. In order to maximize resolution and capture bubble dynamics accurately while reducing computational expense, our simulations use a virtual transducer array, which is a spherical projection of the full transducer array with a much reduced aperture and focal length of $A/12$ and $F/12$. The virtual array, still composed of 18 elements, retains the correct incident wave angles to the stone and is calibrated to account (in an approximate way) for diffraction effects (edge waves) associated with each transducer element.

To calibrate this virtual array, we first ran a large simulation with a full 18-element transducer with focal length $F = 150\,\text{mm}$ and aperture $A = 180\,\text{mm}$ firing at $375\,\text{kHz}$. This simulation was calibrated to achieve a peak negative focal pressure of $-7\,\text{MPa}$. Thinking of the transducer as two concentric rings, there are four edge waves to consider: one traveling inward and one traveling outward radially from each concentric ring. Figure 5.2 shows this for a 2D example, with each of the edge waves labeled 1 through 4. Wave 4 propagates out of the domain, and thus needn't be considered. What remains is to understand the effects of the first 3 waves. We hypothesized that waves 1 and 3, as they propagate inward, would create a focusing effect and increase the pressure near the centerline between transducer and stone center. Conversely, as wave 2 propagates away radially, we conjectured that this would deflect waves from the outer ring of the transducer away from the stone. With a smaller virtual array, these edge waves have far less time to develop before the ultrasound reaches the stone, so we reconstruct these effects through calibration of the input pressure of each transducer element.

We split the virtual transducer into its inner and outer ring, and varied the pressure amplitude for each ring. We gradually increased the amplitude for the six inner elements, and decreased that of the twelve outer elements, while maintaining a peak focal pressure of $-7\,\text{MPa}$ through trial and error. At a pressure amplitude ratio of approximately 1.7 : 1 from outer-ring to inner-ring elements, we reached good agreement between probed pressures across the stone region. Figure 5.3 shows the agreement between the probed pressures at three locations corresponding to the

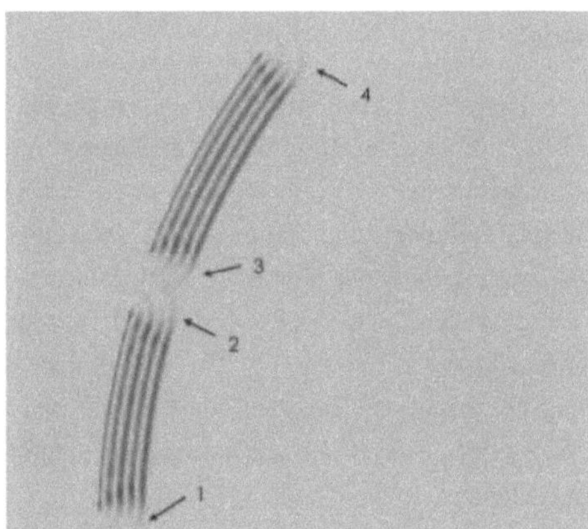

Figure 5.2: 2D example of the 4 edge waves generated by transducer elements, for a transducer with focal length 150 mm **and aperture** 180 mm **with 4 elements. Half of the domain is shown (positive** y **half-domain).**

stone center, edge, and center of the proximal surface for the full and reduced-size transducer for 5 cycles of a 375kHz BWL pulse.

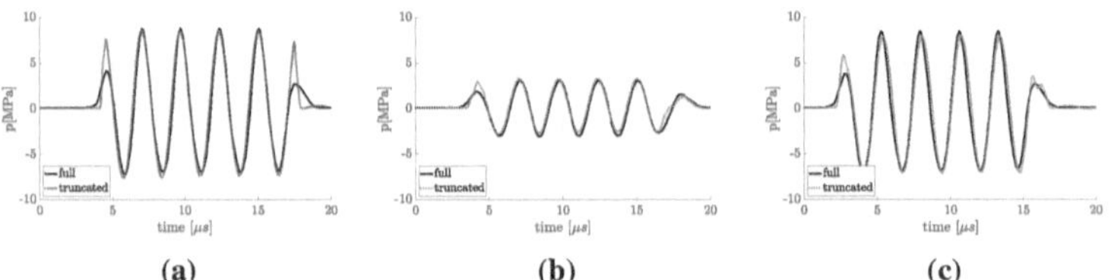

(a)   (b)   (c)

Figure 5.3: Comparison of probed pressures for the full transducer (black solid line) and the scaled-down transducer used in simulations (dotted colored line) at three locations corresponding to the stone center (a), stone edge (b), and center of the proximal stone surface (c).

We note, for clarity, that these simulations are run without a stone, but use the coordinates of these locations on the stone (when present) to probe the pressure. Good agreement is observed between the waveforms over the 5 cycles. While this represents a small difference over the 20-cycle ultrasound pulse used, we do see that the positive focal pressure peaks for the leading and trailing cycles are overestimated by the scaled-down transducer. The resulting scaled-down 18-element virtual array used in simulations is shown in figure 5.4, including in the context of a full BWL simulation in figure 5.4b.

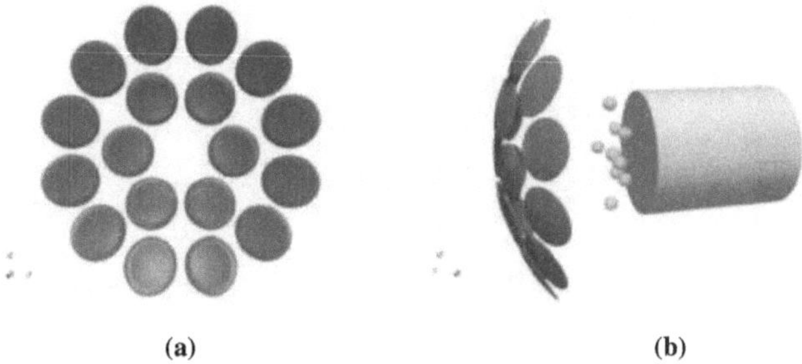

(a)   (b)

Figure 5.4: Virtual array used in BWL simulations viewed face-on (a), and from the side in the context of the full setup of a BWL simulation (b).

5.3 Bubble Shielding during Burst-Wave Lithotripsy

In this section, we utilize a 20-cycle pulse at 375 kHz and compare the results with and without bubbles, in order to corroborate past evidence on shielding while quantifying the effect of bubbles on wave propagation (and damage) within the stone. The simulations shown in this section are 3D, quarter-domain simulations with symmetric boundary conditions in the y and z-directions. As a measure of the performance of our GPU-accelerated numerical framework described in section 4.1.4 of Chapter 4, running on 64 (NVIDIA A40) GPUs on the Delta supercomputer at the National Center for Supercomputing, such a simulation takes approximately 20 hours to run.

Figure 5.5 shows the propagation of waves in the fluid and stone at various times during the simulation. Pressure is plotted in the liquid along a slice of the domain in the y-normal direction, whereas the maximum principal stress σ_1 and continuum damage field D are shown in the top and bottom halves of the stone, respectively. The final damage state in the stone can be seen in figure 5.5d, with the maximum value over time of the maximum principal tensile stress $\text{Max}(\sigma_1)$ in the top half of the stone, and the final damage field D in the bottom half.

We see that in the absence of bubbles, high stress magnitudes are seen across large areas of the stone, and we see two regions of high damage near the stone centerline.

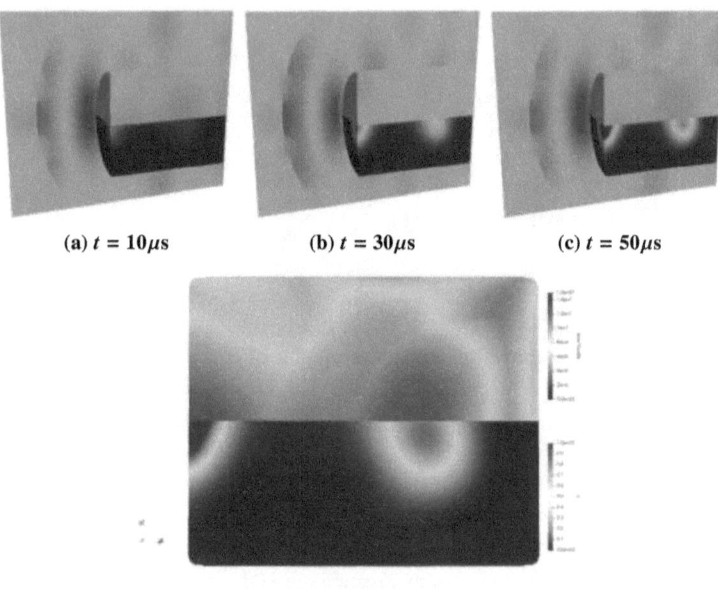

(a) $t = 10\mu s$ (b) $t = 30\mu s$ (c) $t = 50\mu s$

(d) Cumulative

Figure 5.5: Propagation of the maximum principal stress in the stone and damage caused during BWL at 375kHz when no bubbles are present. Figures (a), (b), (c) show three different time steps, and figure (d) shows maximum value of σ_1 in the top half, and damage field D in the bottom half at the end of the simulation.

We now introduce a bubble cloud with a projected area of 3 mm^2 in front of the stone (which we will refer to as the 'dense' bubble cloud), with the closest bubbles at a standoff distance 250 μm, and bubbles randomly distributed along the entire radius of the stone. The resulting maximum principal stress and damage in the stone, as well as bubble cloud oscillations, are shown in figure 5.6. Figure 5.6d shows that the presence of a bubble cloud influences the damage magnitude and pattern in the stone, as acoustic energy can no longer be effectively delivered to the center of the stone. The magnitude of maximum principal stress and damage to the stone is reduced by 16 % and 83 %, respectively (when averaged over the stone volume). We observe slight deformation in the bubbles ahead of the stone over the course of the pulse, but they remain mostly intact.

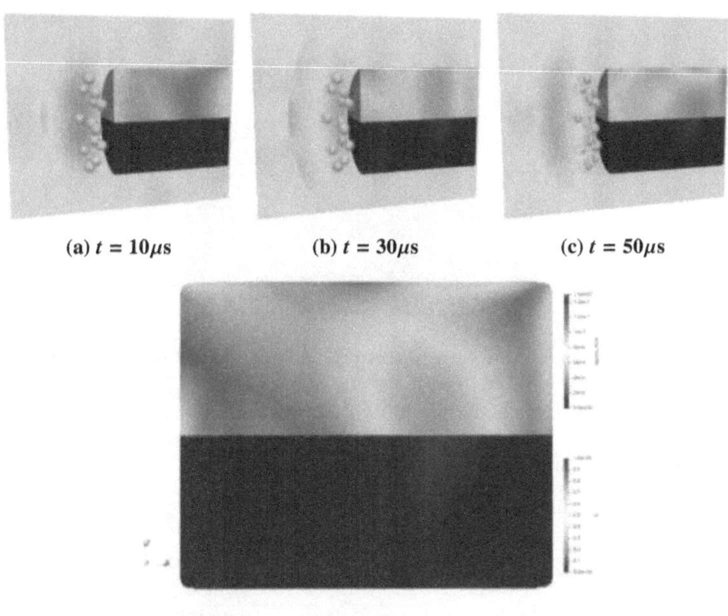

(a) $t = 10\mu s$ (b) $t = 30\mu s$ (c) $t = 50\mu s$

(d) Cumulative

Figure 5.6: Propagation of the maximum principal stress in the stone and damage caused during BWL at 375kHz with a dense bubble cloud ($A = 3\,\mathrm{mm}^2$) present. Bubbles are shown with 2 contour levels for the volume fraction of air: 0.2 (low opacity) and 0.8 (higher opacity). Figures (a), (b), (b) show three different time steps, and figure (d) shows maximum value of σ_1 in the top half, and damage field D in the bottom half at the end of the simulation.

To determine the effect that bubble cloud void fraction has on the stress and damage to the stone, we simulated a second test case with a sparser bubble cloud of projected area $2\,\mathrm{mm}^2$. Figure 5.7 shows the final damage state in this case. The pattern is similar to that seen in the previous case in figure 5.6d, though as expected the stress magnitude and final damage are larger.

To quantify the shielding effects of these bubble clouds, we compare the mean damage in the stone $\frac{1}{V}\iiint_V D$ and mean maximum value of σ_1 over the course of the simulation: $\frac{1}{V}\iiint_V \max(\sigma_1)$, where V is the stone volume. Values are reported in table 5.1. We can see the expected trend, with far higher damage occurring in

Figure 5.7: Damage to the stone during BWL with default transducer parameters and a sparser bubble cloud ($A = 2\,\text{mm}^2$) present. Maximum value of σ_1 over time is plotted in the top half of the stone, and damage field D in the bottom half.

the case with no bubbles present, and an increase in shielding as bubble cloud void fraction increases.

Case	$\frac{1}{V}\iiint_V D$	$\frac{1}{V}\iiint_V \max(\sigma_1)$
No Bubbles	0.052	7.45 MPa
Sparse bubble cloud	0.032	7.18 MPa
Dense bubble cloud	0.009	6.28 MPa

Table 5.1: Comparison of resulting average damage and $\max(\sigma_1)$ values over the stone region for three BWL cases with: no bubbles, a sparse bubble cloud ($A = 2\,\text{mm}^2$) and a dense bubble cloud ($A = 3\,\text{mm}^2$). All cases are with 375 kHz ultrasound.

5.4   Simulations of Bubble Dynamics during BWL

In this section, we aim to understand frequency-typical bubble dynamics in BWL. We begin with two simplified cases to better understand expected behavior of bubbles during treatment. In the first, a simple 1D spherical bubble dynamics model is used, similar to the bubble dynamics model used with DA-IMR in Chapters 2 and 3, to get an idea of the frequency-dependence of bubble dynamics for bubbles of sizes typical of BWL. Next, we will use full 3D simulations of plane ultrasound waves interacting with bubbles and bubble clouds to test hypotheses given by the 1D model. Finally, we conclude this section by using the virtual transducer array described in section 5.2 to model the interactions of spherically focused waves from a multi-element transducer with bubble clouds typical of BWL.

5.4.1   1D Keller-Miksis Model

In the cases simulated in section 5.3, we saw only small oscillations in the bubbles as the ultrasound waves traveled through the cloud. Here, we explore different modifications to the BWL waveform that affect the bubble dynamics more significantly, hoping to collapse or shrink the bubbles, and thus increase damage to the stone by reducing the void fraction of the bubble cloud. Ideally, sufficiently violent collapse of bubbles can be induced, leading to breakup into smaller bubbles and ultimately their dissolution. Given the size of the bubbles ($\approx 250\mu$m), a lower ultrasound frequency will be needed to observe large oscillations in the bubble radius over the time scale of a BWL pulse. To remain within the capabilities of ultrasound transducers, however, we cannot reduce the frequency below $O(10\,\text{Hz})$. As a first approximation of expected bubble dynamics, we implement a solver for spherical bubble dynamics using the Keller-Miksis equation (Keller and Miksis, 1980; Brennen, 1995; Spratt, Rodriguez, Schmidmayer, et al., 2021), which accounts for an incident plane wave with amplitude P and angular frequency ω:

$$
\begin{aligned}
\ddot{R}\left(\frac{4\mu}{\rho} - R(\dot{R} - c)\right) = \frac{1}{2}\dot{R}^3 + \dot{R}\Delta(R) - c\left(\frac{3}{2}\dot{R}^2 + \frac{4\mu\dot{R}}{\rho R} + \frac{2s}{\rho R} - \Delta(R)\right) \\
+ R\dot{R}\Delta'(R) + \left(1 + \frac{\dot{R}}{c}\right)\frac{Pc}{\rho}\sin(\omega)\left(t + \frac{R}{c}\right),
\end{aligned}
\tag{5.1}
$$

where R is the bubble radius, c the sound speed and s the surface tension. Δ is defined as $\Delta(R) = \frac{1}{\rho}(p_b(R) - p_\infty)$, with p_b the bubble pressure assumed uniform, and p_∞ the freestream pressure. We use this model to simulate the response of a $250\,\mu$m bubble to several ultrasound frequencies, ranging from 375 kHz to 20 kHz. The bubble radius time history over 30 µs for representative frequencies is shown in figure 5.8.

Here, we see that while the therapy frequency of 375 kHz only causes very small oscillations of the bubble, lower frequencies induce a more violent collapse. A value of 20 kHz is found to be effective at collapsing bubbles around 250 µm over the course of a BWL pulse, while remaining high enough to be within the range of existing ultrasound transducers. Thus, this frequency is chosen for our low-frequency 'bubble collapsing' wave. In section 5.4.2, we turn to fully 3D plane-

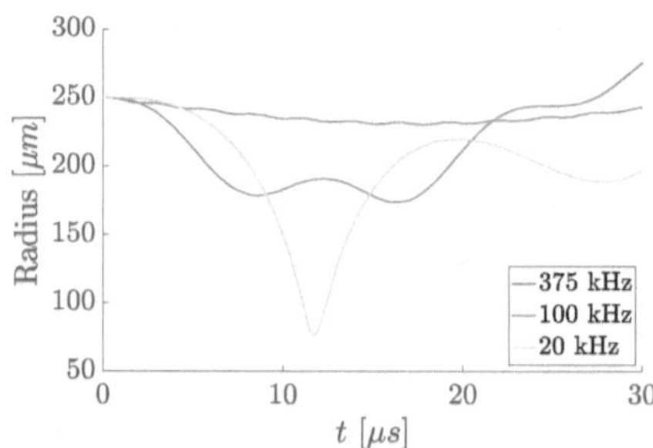

Figure 5.8: Radius time histories for 250 µm bubble under ultrasound at various frequencies with an amplitude of 1 MPa.

wave–bubble simulations to confirm whether this result obtained under a spherical bubble assumption holds for a fully resolved 3D bubble.

5.4.2   3D Plane-Wave–Bubble Simulations

We first compare the oscillations of a single bubble under a plane wave at 375 kHz and 20 kHz with a magnitude of 3 MPa. This is shown in figures 5.9 and 5.10. Different bubble dynamics are observed. As expected, the low frequency wave collapses the bubble early in the simulation, which rebounds but remains small throughout. By contrast, the bubble radius oscillates only slightly when exposed to the 375 kHz ultrasound. This matches what was expected based on the Keller-Miksis model simulations and the radius curves shown in figure 5.8.

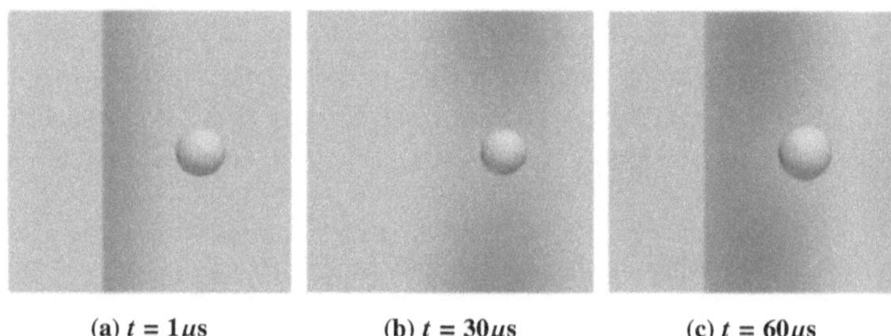

(a) $t = 1\mu$s　　　　(b) $t = 30\mu$s　　　　(c) $t = 60\mu$s

Figure 5.9: Oscillation of a 250 µm bubble under a plane wave at 375 kHz.

We also simulate plane waves interacting with the dense bubble cloud used in section 5.3. The dynamics of the bubble cloud are such that there is some deformation of

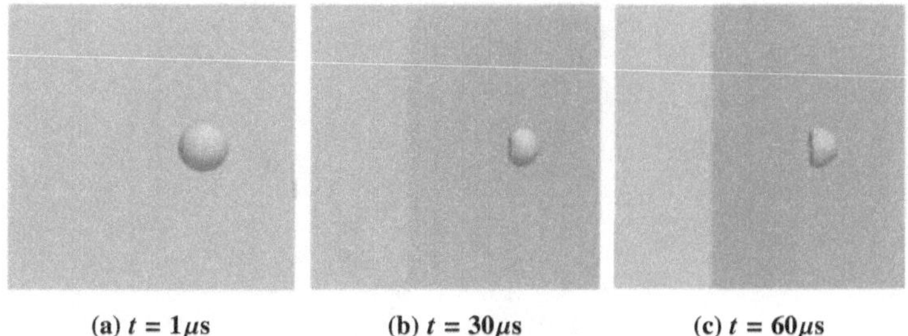

(a) $t = 1\mu$s   (b) $t = 30\mu$s   (c) $t = 60\mu$s

Figure 5.10: Oscillation of a 250 μm bubble under a plane wave at 20 kHz.

the bubbles in both cases, but the overall results remain that the low frequency wave more effectively collapses and deforms bubbles in the cloud. Figure 5.11 shows the behavior of a bubble cloud under 375 kHz ultrasound. In Figure 5.12, a 20 kHz pulse of equal amplitude is superposed on the 375 kHz pulse. This causes large deformation in the bubble cloud, with many bubbles largely collapsed by the end of the simulation in the dual-frequency case, with a contour level of $\alpha^{(air)} = 0.9$

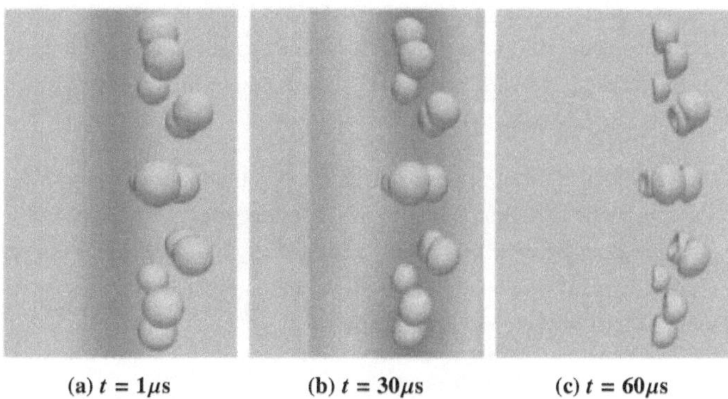

(a) $t = 1\mu$s   (b) $t = 30\mu$s   (c) $t = 60\mu$s

Figure 5.11: Oscillation of a dense bubble cloud ($A = 3\,\text{mm}^2$) under a plane wave at 375 kHz.

This shows that adding a low frequency 'bubble collapsing pulse' to the usual 375 kHz 'therapy pulse' may help reduce shielding by collapsing bubbles and thus reduce the void fraction of the cloud. While this appears to be effective, further study of cloud bubble dynamics under ultrasound pulses could help formulate other strategies. Given the high dependence of bubble dynamics on bubble radius and characteristics of bubble clouds, determining a systematically optimal frequency

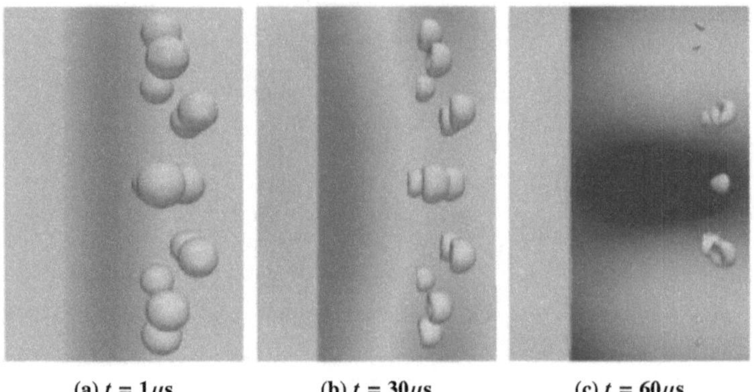

(a) $t = 1\mu$s   (b) $t = 30\mu$s   (c) $t = 60\mu$s

Figure 5.12: Oscillation of a dense bubble cloud ($A = 3$ mm^2) under the superposition of plane waves at 375 kHz and 20 kHz.

may prove difficult, as these characteristics vary case-by-case during BWL. By choosing this low frequency, and setting the phase of the ultrasound so that the pressure pulse begins as positive, we can ensure the fast collapse of bubbles across a range of sizes and for different bubble cloud configurations. Thus, we reduce the need for fine-tuning, and opt for a strategy that should be broadly applicable given various bubble cloud configurations.

With a full 18-element therapy transducer, there are multiple ways this low frequency pulse can be added. A first approach is to simply superpose the 20 kHz sine wave signal to each transducer element. Alternatively, individual elements can be controlled to fire at a single frequency. Both approaches are examined in section 5.4.3 for transducer–bubble-cloud simulations.

5.4.3 Spherically focused Ultrasound–Bubble Cloud Simulations

To examine the dynamics of a bubble cloud in an acoustic field akin to that of BWL, the simulations presented in this section make use of the virtual array described in section 5.2. In all these simulations, the denser bubble cloud with projected area 3 mm^2 is used, where the interaction between bubbles in the cloud has a larger effect on its dynamics. Three cases are shown. First, a reference simulation with only the 375 kHz ultrasound is shown in figure 5.13.

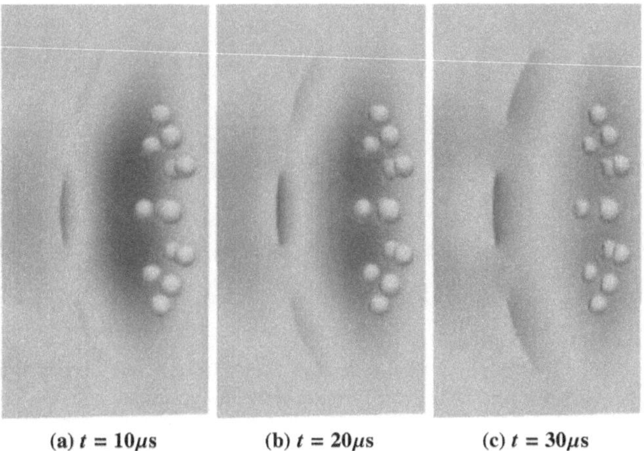

(a) $t = 10\mu s$ (b) $t = 20\mu s$ (c) $t = 30\mu s$

Figure 5.13: Oscillation of a dense bubble cloud ($A = 3\,\text{mm}^2$) exposed to BWL transducer ultrasound at 375 kHz.

Then, a case where a 20 kHz ultrasound with a focal amplitude of 1 MPa is superposed on each transducer element, which we will call the 'dual frequency configuration,' is shown in figure 5.14. Finally, a simulation where the 6 inner ring elements fire the 20 kHz pulse, while the outer ring fires at the usual 375 kHz, referred to as the 'split frequency configuration,' is shown in figure 5.15. Figure 5.16 shows this split between inner ring and outer ring, with elements colored in orange at 20 kHz and those in blue at 375 kHz. In all tested cases, focal pressure was calibrated not to exceed a peak of 7 MPa to remain within a safe treatment regime.

The first thing to note is the similarity between the bubble cloud dynamics in the plane-wave and transducer cases, apparent when comparing figures 5.13 and 5.14 to the results of section 5.4.2. The notable difference being that the cloud collapses faster under the incident spherical wave, mostly due to higher pressures near the focus of the transducer. The deformation of the bubbles in the dual and split-frequency cases is similar, though the final bubble cloud shape is slightly different. In the dual-frequency case, the cloud is slightly more compact and concave to the right, while it is more spread out and concave to the left in the split-frequency case. While this appears to be a minor difference, we will see in section 5.6 that it has an impact on the shielding of the cloud.

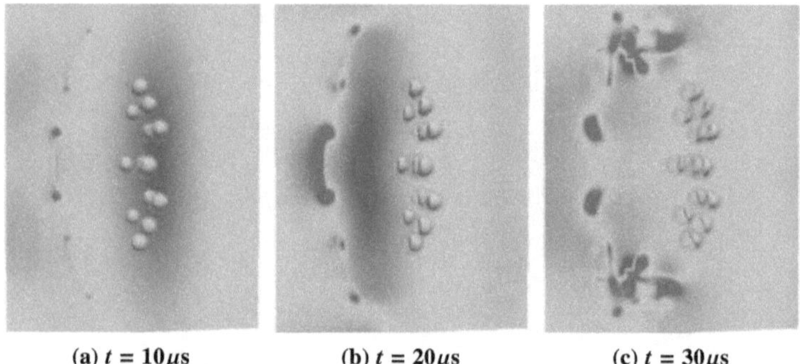

(a) $t = 10\mu s$ (b) $t = 20\mu s$ (c) $t = 30\mu s$

Figure 5.14: Oscillation of a dense bubble cloud ($A = 3\,\text{mm}^2$) exposed to BWL transducer ultrasound in the dual frequency configuration.

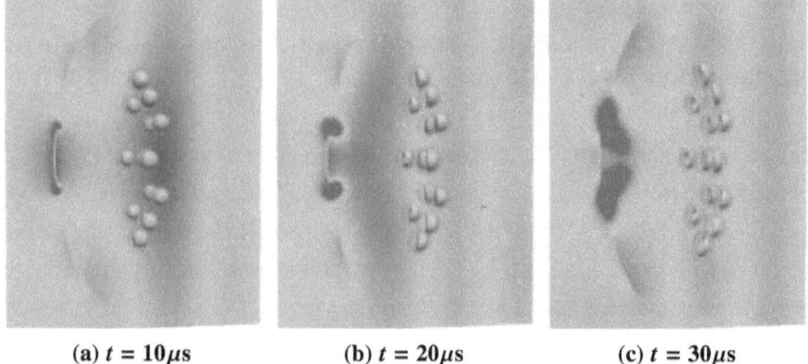

(a) $t = 10\mu s$ (b) $t = 20\mu s$ (c) $t = 30\mu s$

Figure 5.15: Oscillation of a dense bubble cloud ($A = 3\,\text{mm}^2$) exposed to BWL transducer ultrasound in the split frequency configuration.

Overall, both the split-frequency and dual-frequency configurations are effective at collapsing the bubble clouds quickly during the course of the ultrasound pulse. From these transducer–bubble cloud simulations alone, it appears that both strategies could be viable to increase the damage done to the stone in BWL in the presence of a bubble cloud. In the following section 5.5, we examine both strategies in the context of a full transducer–bubble-cloud–stone simulation, and compare the damage to the stone with each configuration.

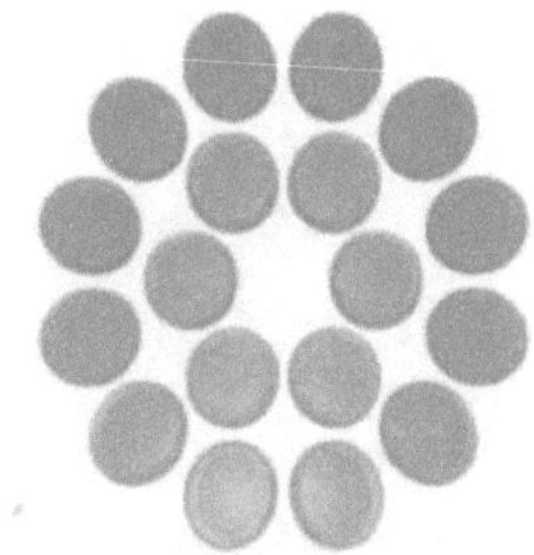

Figure 5.16: Split frequency transducer configuration: orange elements fire at 20 kHz, blue elements at 375 kHz.

5.5 Dual and Split-Frequency BWL Simulations

Adding the stone back into the simulations, we now carry out full 3D simulations of BWL with the two strategies described in the previous section. We begin by looking at the dual-frequency configuration. While there may be similarities with the bubble cloud behavior observed in section 5.4.3, the reflection of acoustic waves off of the surface of the stone impacts bubble cloud dynamics. Figure 5.17 shows the propagation of σ_1 and damage state D throughout the simulation, as well as the evolution of the bubble cloud in the dual-frequency case. We see that as with the simulation shown in figure 5.14, bubbles are significantly deformed, though bubbles deform and collapse faster in this case due to the higher pressure magnitudes experienced in the bubble cloud due to reflected waves off the stone.

However, as seen in figure 5.17d which shows the final damage state, this configuration does not improve damage to the stone compared to the case with only the 375 kHz wave shown in figure 5.6d. With this transducer configuration, the bubble cloud continues to effectively shield the stone, and it is thus not an effective method to reduce shielding.

In contrast, we now turn to the split-frequency configuration. Figure 5.18 shows a time history of this simulation with a dense bubble cloud. We see that there is again a significant deformation of bubbles and a reduction in void fraction. In this case, though, the bubbles are stretched and pushed out radially, away from the center of the proximal stone surface. This difference in bubble cloud dynamics appears to

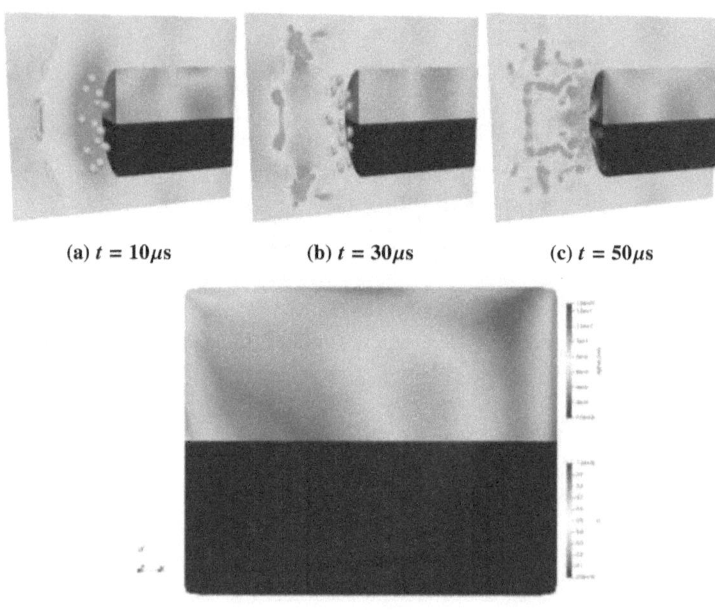

(a) $t = 10\mu s$ (b) $t = 30\mu s$ (c) $t = 50\mu s$

(d) Cumulative

Figure 5.17: **Propagation of the maximum principal stress in the stone and damage caused during BWL with the dual-frequency configuration with a dense bubble cloud ($A = 3\,\mathrm{mm}^2$) present. Bubbles are shown with 2 contour levels for the volume fraction of air: 0.2 (low opacity) and 0.8 (higher opacity). Figures (a), (b), (c) show three different time steps, and figure (d) shows maximum value of σ_1 in the top half, and damage field D in the bottom half at the end of the simulation.**

have an important effect on stone shielding, as the damage to the stone is much more significant in this case.

Indeed, figure 5.18d shows the high damage and large maximum magnitude of σ_1 observed in this case. We also simulated this configuration with the sparse bubble cloud, with similar results apart from minor differences in the damage patterns. The split-frequency configuration is far more effective than the dual-frequency case, causing high damage and stress magnitudes in the stone.

Table 5.2 compares average damage field value D and maximum σ_1 in the stone over the course of the simulation in the case with only the 375 kHz ultrasound and

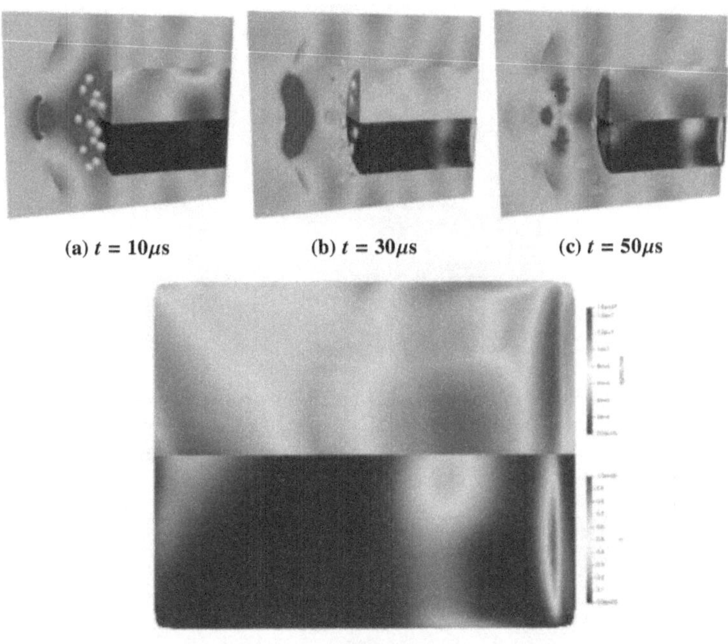

(a) $t = 10\mu s$ (b) $t = 30\mu s$ (c) $t = 50\mu s$

(d) Cumulative

Figure 5.18: Propagation of the maximum principal stress in the stone and damage caused during BWL with the split-frequency configuration with a dense bubble cloud ($A = 3\,\mathrm{mm}^2$) present. Bubbles are shown with 2 contour levels for the volume fraction of air: 0.2 (low opacity) and 0.8 (higher opacity). Figures (a), (b), (c) show three different time steps, and figure (d) shows maximum value of σ_1 in the top half, and damage field D in the bottom half at the end of the simulation.

the split-frequency case. In both cases, the split frequency significantly improves damage to the stone. This is particularly effective with the dense bubble cloud, where we observe a nine-fold increase in the mean damage value in the stone, and a 25% increase in the mean value of $\max(\sigma_1)$. This method thus appears to be highly effective at reducing bubble cloud shielding in both tested cases.

Figure 5.19: Damage to the stone during BWL with split-frequency configuration and a sparse bubble cloud ($A = 2\,\mathrm{mm}^2$) present. Maximum value of σ_1 over time is plotted in the top half of the stone, and damage field D in the bottom half.

Case	wave form	$\frac{1}{V}\iiint_V D$	$\frac{1}{V}\iiint_V \max(\sigma_1)$
Sparse bubble cloud	BWL at 375 kHz	0.032	7.18 MPa
	split-frequency	0.080	7.95 MPa
Dense bubble cloud	BWL at 375 kHz	0.009	6.28 MPa
	split-frequency	0.082	7.84 MPa

Table 5.2: Comparison of resulting average damage and $\max(\sigma_1)$ values over the stone region between the baseline BWL 375 kHz waveform case and dual frequency case for a sparse bubble cloud ($A = 2\,\mathrm{mm}^2$) and a dense bubble cloud ($A = 3\,\mathrm{mm}^2$).

5.6 Discussion

For both the dual-frequency and split-frequency configurations, bubble cloud void fraction reduction occurs on the same time scale, as shown in sections 5.4.3 and 5.5. There is a difference, however, in the dynamics and deformation of the bubbles. It appears that the split frequency configuration causes favorable conditions for the propagation of acoustic waves through the bubble cloud. In figure 5.20, we show a side view of the bubble clouds from the dual-frequency case and split-frequency cases after 30 µs (from the same simulations as figures 5.17 and 5.18).

Here we see that the shapes of the bubble clouds are quite different. Figure 5.20a shows a more compact bubble cloud which is slightly concave to the right, while the bubble cloud in figure 5.20b is concave to the left and more spread out in the y-z

 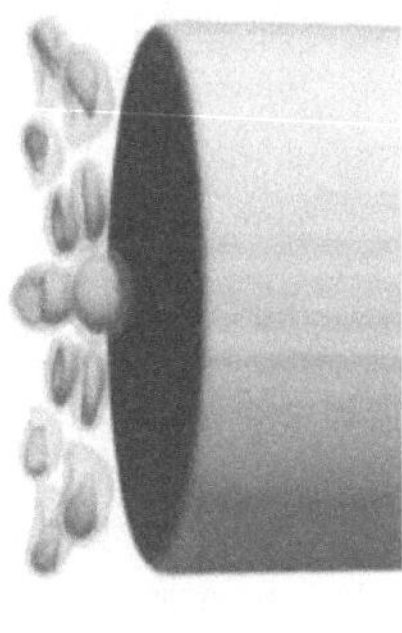

(a)   (b)

Figure 5.20: Dense bubble cloud ($A = 3\,\mathrm{mm}^2$) deformation during treatment at $t = 30\mu s$ for the (a) dual frequency and (b) split frequency cases. Bubbles are shown with 2 contour levels for the volume fraction of air: 0.2 (low opacity) and 0.8 (higher opacity)

plane, with the edges of the cloud extending beyond the stone edge. This difference in shape can be explained by the fact that the bubbles are being pushed out by the low frequency waves. In the dual frequency case, these low-frequency waves are coming from the entire transducer, pushing the outer bubbles inward towards the center. By contrast, the narrower beam of low-frequency waves in the split-frequency case pushes bubbles at the center of the cloud towards the stone, and as these bubbles interact with the stone surface they are pushed out radially, causing the cloud to spread out.

In fact, this effect was visible in simulations from section 5.4.3, in figures 5.14c and 5.15c, even without the stone present. The bubble cloud in the former is more compact, and in the latter spreads out radially. Given these examples, it appears that over a single ultrasound pulse, a strategy that spreads a bubble cloud out may be more effective than one aimed simply at collapsing bubbles or reducing their void fraction.

We note here that these results are limited to the time-scale of a single BWL pulse. Even with a fast pulse repetition frequency, the time between pulses is three orders of magnitude larger than the time-span of a BWL pulse (0.1 s vs. 60 µs). While this method of spreading out the bubble cloud is effective at causing high damage to the stone over a single pulse, the resulting bubble dynamics between pulses are not yet well understood. Intuitively, spreading bubbles away from the stone as is done with the split-frequency should only improve the energy delivery to the stone in the subsequent pulse, and so the effects of this strategy could stack up with each subsequent pulse. However, this remains to be tested, and is beyond the capabilities of our numerical methods.

Finally, in a clinical setting, the focal point accuracy required for the transducer may be difficult to achieve. The schism between damage caused in the split-frequency and dual-frequency cases may not be as obvious in a case where the transducer focal point is not exactly in the center of the stone. With low frequency wave in the center of the beamwidth, however, the tendency should remain for bubbles to spread out and away from the focal point of the transducer, and thus retain the benefits observed above even with imperfect focusing and a stone with a more complex topology.

Overall, the presented strategy in the split-frequency configuration appears to be a promising approach to reduce the shielding of bubble clouds by collapsing and dispersing bubbles forming ahead of the stone. Both the dense and sparse bubble clouds presented were effective in shielding the stone and reducing damage. However, in the tested cases, the split-frequency transducer configuration improved damage to the stone. It is a promising way to improve the efficacy and efficiency of BWL in cases with bubbles present ahead of the stone.

Chapter 6

CONCLUSIONS AND FUTURE WORK

The goal of this book was to implement and apply numerical models and data-driven methods to study biological systems involving the interaction of cavitating gas bubbles with elastic and viscoelastic materials. We successfully developed a data assimilation framework (DA-IMR) to aid in the modeling of biomaterials in low strain, high strain-rate settings, and applied it to various experimental data sets with distinct materials, bubble nucleation methods, and viscoelastic material models. DA-IMR was thoroughly tested for soft material characterization, and has improved both practicality (from O(days) to O(minutes)) and accuracy of the existing framework. We have shown its usage in a broad array of cases, and its ability to adapt to varied material models or additional parameters to estimate, with negligible added computational cost. Additionally, its use helped further our understanding of the limitations of current physical models used in inertial microcavitation rheometry. We have also implemented a numerical framework for the direct numerical simulation of acoustic-wave–bubble–elastic solid interactions, enabling the full simulation of burst wave lithotripsy. Through these simulations, we have investigated bubble dynamics and damage occurring in stones during BWL, and proposed strategies to mitigate bubble shielding, thus improving damage delivered to kidney stones, and accelerating their comminution. Key conclusions from each chapter of this book and suggestions for future work are summarized below

6.1 Data Assimilation for Inertial Microcavitation Rheometry

In Chapter 2, we presented a data-assimilation approach for the characterization of viscoelastic materials via observation of bubble collapse. This extension of inertial microcavitation rheometry, which we call DA-IMR, accurately estimated the mechanical properties of soft materials using ensemble-based data assimilation methods. With surrogate data obtained by adding random noise to simulated bubble radius vs. time curves (at frame rates replicating available high-speed imaging equipment), we validated this approach and demonstrated its accuracy and reliability in the context of laser-induced cavitation in hydrogels. The iterative ensemble Kalman smoother (IEnKS) and a hybrid ensemble-based 4D-Var method (En4D-Var) provided the best results. The En4D-Var reduced computational cost by up to two orders of magnitude, obtaining accurate parameter estimates efficiently, accurately, and in a scalable framework. The IEnKS, while less efficient, provided useful information about time-dependent modeling uncertainty.

In Chapter 3, the DA-IMR framework was applied to experimental data across three test cases. In the first, it was used with experimental data similar to that used in the validation case in Chapter 2. Here, it was shown to provide accurate estimates for shear modulus and viscosity of a polyacrylamide gel with experimental laser-induced cavitation data, despite experimental inconsistency, model error, and noisy data. Additionally, we gained insight into data sets where the spherical bubble dynamics model used may have inadequately captured physics during violent bubble collapse. Next, we applied DA-IMR to experimental data for laser-nucleated cavitation in hydrogels with seeded particles. Here, a different stretch ratio regime was reached by nucleating bubbles on the surface of micro-particles in the gels. In this regime, and using a more general viscoelastic material model, we showed the adaptability and versatility of DA-IMR. At the same time, we established shortcomings in the physical model near the bubble collapse point in cases with high laser energy, as revealed by the quasi-online IEnKS estimation. This highlighted model uncertainties linked to plasma formation and the optical breakdown process during laser-induced cavitation. Finally, DA-IMR was used with ultrasound-induced cavitation, which addresses the issues seen previously with laser-induced cavitation. Here, the En4D-Var again accurately estimated material properties, using three different material models and estimating two additional model parameters.

Possible Improvements and Future Work The modeling uncertainties seen with laser-induced cavitation, particularly those analyzed in section 3.2, point to the need to capture additional physics in the bubble-dynamics models used. Some of the concerns around modeling uncertainties due to violent bubble collapse were addressed by switching to ultrasound-nucleated cavitation. However, with a better understanding of the plasma physics and rapid collapse occurring following laser-induced cavitation, this could become a more accurate and dependable framework for material characterization. Extending DA-IMR to more complete models, including those which could account for non-sphericity of bubbles, could also expand the capabilities of this framework. While such models are likely to be far more computationally expensive, the gained efficiency and parallelization capability of

DA-IMR suggest that such an approach would be computationally feasible. Such an adaptation could help gain insight into the non-spherical collapse of bubbles in viscoelastic materials, beyond the scope of material characterization.

While a significant advantage of DA-IMR is its efficiency and scalability due to the parallelizable nature of ensemble-based methods, this efficiency can be reduced given the need to tune algorithm parameters. In these initial runs, parameters such as initial modeling and measurement error covariance, number of ensemble members, and covariance inflation (both additive and multiplicative) needed to be tested and set through trial and error to obtain optimal results. An approach to tune these parameters based on the statistics of available data could further improve the efficiency of the method.

6.2　A Numerical Framework for Full Simulations of Burst-Wave Lithotripsy

In Chapter 4, we presented our implementation of a framework to model the full interaction of ultrasound waves from a spherically focused transducer array with gas bubbles and a submerged stone. This model was validated, and shown to match experimental stress imaging results during BWL well. Furthermore, the GPU implementation of the code was shown to provide significant acceleration and near-ideal scaling, enabling high resolution simulations. Calculating the maximum principal stresses generated in stones of various shapes and sizes, we confirmed that a frequency range of 350 kHz to 375 kHz appears effective in breaking kidney stones of typical size. That is, it maximizes the stresses and damage generated in stones of various sizes typical of human renal calculi.

In Chapter 5, we used the framework presented in Chapter 4 to investigate bubble dynamics and damage in burst-wave lithotripsy. We quantified the shielding of bubble clouds on waves and damage in the stone. We determined that introducing a low-frequency 20 kHz wave in addition to the usual BWL frequency, in this case 375 kHz, can effectively collapse bubbles ahead of a kidney stone during treatment. When splitting the therapy transducer between a low-frequency inner ring and high-frequency outer ring, shielding due to the tested bubble clouds was effectively reduced. We saw that the manner in which the low-frequency wave is introduced is important to maximize damage, and that bubble dispersion plays a role in this shielding reduction. We have presented a transducer configuration that enabled high damage in the tested cases, with a low-frequency pulse in the center of the transducer to disperse bubbles, and high-frequency pulse from the outer ring to deliver damage to the stone.

Perspectives on Future Applications Only a few strategies to minimize bubble shielding have been tested here. Many other approaches exist, be it with the 18-element transducer used here or other transducer and acoustic source configurations. There are likely other ways to reduce shielding, and other approaches focused on pushing bubbles away from the stone, may yield better results. Furthermore, given the vast difference in time scales between BWL pules, and inter-pulse time, only single pulses can be simulated with this method. Experimental trials or reduced order modeling methods that can cover the time-scale of multiple successive pulses could shed light on how well this method would translate to a therapy setting.

While multi-pulse simulations remain a challenge, the far reduced computational cost afforded by the GPU implementation broadens how this method can be applied. Indeed, iterative optimization approaches are possible with relatively high resolutions, even in full 3D simulations, provided the domain is small enough. Given the possibility of simulating damage in any arbitrary stones, one could envisage a treatment protocol whereby a stone is first imaged (using ultrasound or other imaging techniques), and an optimization is performed given this stone's size and shape using MFC. In simulations without bubbles present, the necessary resolution to obtain accurate stress and damage outputs is low, and such optimizations could thus readily be performed.